Biodiversity of Mediterranean Ecosystems in Australia

Heathlands in the mediterranean regions of Australia, particularly in the south-west, are among the most diverse plant communities in the world. (Photo: R. J. Hobbs).

Biodiversity of Mediterranean Ecosystems in Australia

Edited by
Richard J. Hobbs

Printed and published by
Surrey Beatty & Sons Pty Limited

The National Library of Australia
Cataloguing-in-Publication entry:

Biodiversity of Mediterranean Ecosystems in Australia

ISBN 0 949324 49 3.

1. Biological diversity — Australia.
2. Ecology — Australia.
I. Hobbs, R. J. (Richard J.).

574.50994.

Published September 1992

PRINTED AND PUBLISHED IN AUSTRALIA BY
SURREY BEATTY & SONS PTY LIMITED
43 Rickard Road, Chipping Norton, NSW 2170

Preface

Biodiversity has emerged as one of the key concepts of conservation, receiving the attention not just of scientists and environmentalists, but also of politicians and the general public. National and international strategies have been and are being developed for the conservation of biodiversity. Despite this, the concept remains somewhat woolly, and there is considerable discussion over the reasons for conserving biotic diversity and the urgency with which the problem must be tackled. A variety of reasons can be put forward for the conservation of biodiversity, but a commonly cited reason is that the maintenance of diversity is essential for the continued existence of natural ecosystems, and hence for the continued supply of the essential services they supply the ever-growing human population.

Despite the rhetoric, there is little scientific evidence to suggest that biodiversity is an important feature in maintaining functioning ecosystems, mainly because scientific studies have rarely been designed to examine this hypothesis. In recognition of this, a world-wide scientific programme was initiated in 1991 to examine the role of biodiversity in ecosystem functioning. Under the auspices of the International Union of Biological Sciences (IUBS) and the Scientific Committee on Problems in the Environment (SCOPE), the programme aimed to stimulate debate and investigation of the question. A number of foci were chosen for the programme, including ecosystem types of particular interest and specific problem areas. Mediterranean-type ecosystems were chosen as one of the special foci, and each region containing mediterranean-type ecosystems (the Mediterranean basin, California, Chile, South Africa and Australia) was to start the process by synthesizing available information for later inclusion in an overall mediterranean ecosystem perspective. I was asked to co-ordinate the Australian contribution, and accordingly I organized a two-day workshop in Perth in December 1991. This workshop drew together a small number of scientists with experience in various aspects of mediterannean ecosystems in Australia. The objective of the workshop was not to summarize existing data (partially since the database is remarkably small), but to explore and develop ideas on the role of biodiversity in ecosystem functioning in mediterranean Australia. Using a mixture of individual presentations and group discussions, the group went some way towards clarifying the often confusing concepts and reaching some agreement on how to treat the topic. This book represents the outcome of the group's discussions, and forms the basis for the Australian contribution to the world-wide focus. All the chapters in the book have been rigourously refereed, both by other workshop participants and by other scientists not present at the workshop.

I wish to thank Dr Brian Walker who supported the project and made CSIRO Division of Wildlife and Ecology funds available for running the workshop. I also thank the following for their critical reviews of manuscripts; Graham Arnold, David Goodall, Robert Lambeck, Byron Lamont, Sandra Lavorel, Ted Lefroy, Bill Loneragan, Bert Main, Jonathan Majer, Harry Recher, Denis Saunders, Jeff Short, Mark Stafford-Smith, Brian Walker, Ken Wallace and Ray Wills. Bill Loneragan, Robert Lambeck, Ed Witkowski and Ted Lefroy acted as scribes for group discussions, and Lesley Brooker compiled the index. Thanks also to Lyn Atkins who made sure that lunches were available for workshop participants. I would particularly like to thank the authors of chapters in this book for their efforts in completing manuscripts as quickly as their commitments would allow, so that the publication of this book could be carried out as rapidly as possible. In addition, I am extremely grateful to Ivor Beatty and Surrey Beatty and Sons for agreeing to publish the book and doing this so effectively.

Richard Hobbs

Participants in the Workshop on Ecosystem Function of Biodiversity in Mediterranean Ecosystems of Australia, held at CSIRO, Division of Wildlife and Ecology, Perth, in December 1991.

Left to right: Ed Witkowski, Robert Lambeck, Richard Hobbs, Ted Lefroy, Sandra Lavorel, Jonathan Majer, Richard Groves, Denis Saunders, Bert Main, Byron Lamont and Steve Hopper.

Contents

List of Colour Plates

CHAPTER 1

Function of biodiversity in mediterranean ecosystems in Australia: definitions and background

Richard J. Hobbs[1]

Introduction

The diversity of life on Earth is both fascinating and perplexing — fascinating because of its endless variety, beauty and ability to cope with the Earth's varied environments, and perplexing because the task of documenting and understanding the Earth's biota is so huge and demanding. We do not know how many species there are on the Earth, or much about the distribution, ecology and dynamics of all but a few species. And yet it seems likely that a significant proportion of the Earth's biota could become extinct over the next 50–100 years as a result of human activities (Ehrlich and Ehrlich 1981; Wilson 1988).

Does this impending loss of species matter? In this chapter I examine what is meant by biodiversity, and what the potential values are of species diversity, from a variety of standpoints. In particular, as a precursor to the subject matter of this book, I examine the concept of ecosystem functioning and the potential role that individual species have in that. I then describe the mediterranean ecosystems of Australia and indicate their value as test systems for the study of the ecosystem function of biodiversity.

This book is the result of a workshop held in Perth, Western Australia in December 1991, which was convened as part of a SCOPE programme on the ecosystem function of biodiversity.

---

[1]CSIRO Division of Wildlife and Ecology, LMB 4, PO Midland, Western Australia, Australia 6056.

Pages 1–25 *in* BIODIVERSITY OF MEDITERRANEAN ECOSYSTEMS IN AUSTRALIA ed by Richard J. Hobbs. Surrey Beatty & Sons, 1992.

This programme consists of a number of syntheses over the period 1991–94, covering general principles, priority systems and problem issues. Mediterranean ecosystems were selected as one of the priority foci, and this book represents the Australian contribution to that regional focus. The objective was "to assess and synthesize knowledge and understanding of the role of biodiversity in ecosystem function (i.e., energy, nutrient, water fluxes and other 'emergent' properties), using evidence from mediterranean ecosystems" (B. Huntley, pers. comm.). This book looks first at the patterns of biodiversity present (Chapters 2 and 3), and then considers the effects of changes in land use and climate on biodiversity (Chapter 4). Three differing perspectives on the ecosystem function of biodiversity then follow (Chapters 5–7). Chapter 8 considers various approaches to the study of the function of biodiversity, and Chapters 9 and 10 discuss the maintenance and restoration of diversity, and the management requirements for this. Finally, Chapter 11 provides a synthesis of the information presented in the book, together with recommendations for research and management.

Biodiversity

Definition

Biodiversity is defined as encompassing "Earth's 10 million or so species of plants, animals and other organisms; the genetic variability of each species; and the different ecosystems they form" (WCU/UNEP/WWF 1991). Implicit in this definition is the idea of a variety of levels of organization, meaning that biodiversity does not simply mean numbers of species. Soulé (1991) recognizes a "biospatial hierarchy" of levels, which are also implicit in the research agenda for biodiversity developed early in the SCOPE programme (Solbrig 1991). These levels are:

1. Whole systems at the landscape or ecosystem levels
2. Assemblages (associations and communities)
3. Species
4. Populations
5. Genes

Noss (1990) has gone further to suggest that there is more to the hierarchy than just the compositonal aspects, and he recognizes that structural and functional aspects of biodiversity are also important (Fig. 1).

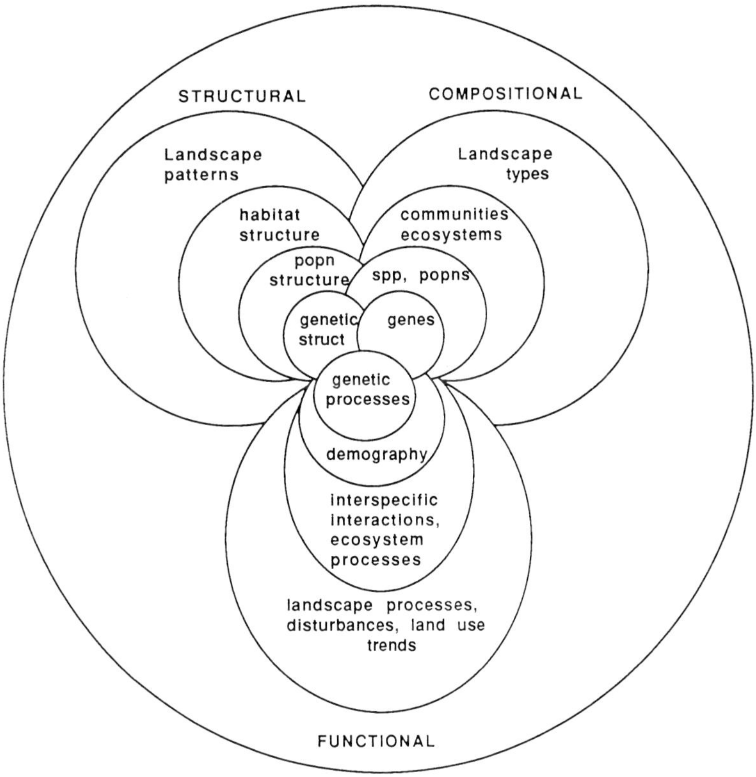

Fig. 1. Multiple levels of organization of biodiversity, illustrating compositional, structural and functional components (after Noss 1990).

While it is relatively easy to define biodiversity in this way, it is less easy to quantify it at any of the levels or to assess the significance of the compositional, structural or functional aspects. The overall biodiversity of the Earth is poorly documented and understood (Mann 1991), and it seems likely that only a small proportion of the actual number of species present have been collected and/or described, especially in tropical regions. This makes predictions about extinctions difficult; how can you predict that a species will go extinct if you don't know it exists? Particular groups of organisms are very poorly documented, especially the insects. Table 1 provides examples of invertebrate groups in Australia which still need extensive taxonomic work. Other diverse and abundant groups, for instance Nematoda, are even less well known. The number of recognized species is thus liable to increase

Table 1. Examples of the current knowledge of invertebrate groups in Australia (after Taylor 1983 and Williams 1990).

	% named species	% known spp. not yet named	Estimated % spp. not yet collected	Estimated total no. spp.
Collembola	10	10	80	2 500
Ephemoptera	10	60	30	300
Orthoptera	27	51	22	2 840
Coleoptera	55	8	37	33 500
Lepidoptera	47	21	30	21 870

as taxonomic and survey work progresses. This apparent increase in species may thus give a confusing impression when contrasted with claims that species extinctions are occurring or imminent.

The study of biodiversity poses many other scientific problems, including the question of how best to measure diversity at the various organizational levels. Spatial components of species diversity provide an example of how complex the situation can become. Recent reviews by Walker (1989) and Bond (1989) discuss the components of species diversity in terms of α, or within habitat diversity, β diversity, or the turnover of species between habitats and γ diversity or the total diversity from all assemblages within a landscape. Also discussed are δ diversity, representing differences between landscapes or habitats in a geographical region and ϵ diversity, or the total for differing landscapes in a large geographical region. Some treatments swap the definitions of γ and δ diversity (Fig. 2; Cody 1975; Bond 1989), emphasizing the lack of rigorous definition in this field. It is clear, however, that the quantification of biodiversity is not a simple task, especially when one also considers the structural and functional components. In this book we will examine the various levels of organization in the preliminary chapters, and thereafter most discussion will center around α, or within habitat diversity. We will, however, consider landscape-level phenomena as appropriate.

Why is biodiversity important?

The diversity of life on Earth is currently subject to a number of threats resulting from human activities. Soulé (1991) considered the most important of these to be:

1. the loss of habitat
2. the fragmentation of habitat
3. overexploitation

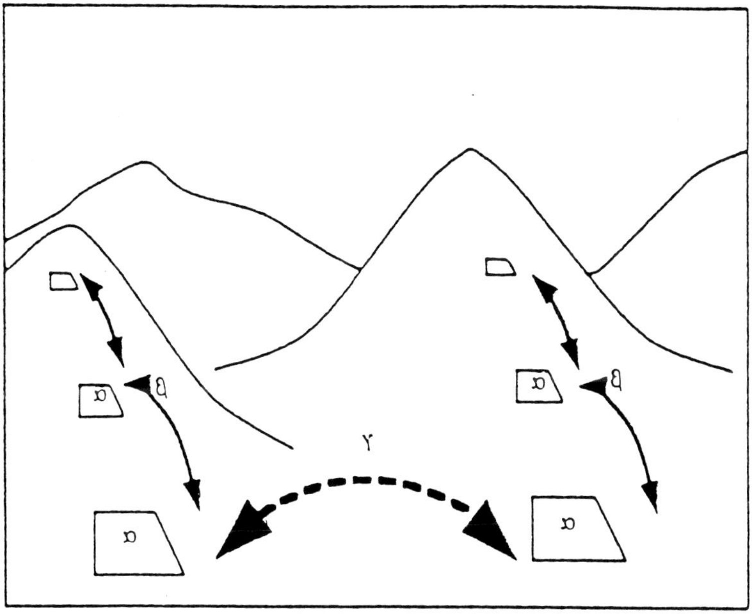

Fig. 2. Three major components of species diversity. α = species richness within a habitat, β = species turnover between habitats, γ = species turnover within habitats between geographic regions (after Bond 1989).

4. spread of exotic species and diseases
5. air, soil and water pollution
6. climate change

Soulé (1991) also points out, however, that although these factors constitute the proximal causes of species loss, the ultimate causes are to be found in the current human condition, with factors such as population growth, poverty, current economic paradigms and human attitudes towards nature being of prime importance.

Is the potential loss of species important? Numerous arguments are put forward in support of conserving biodiversity. In the recent document "Caring for the Earth: a Strategy for Sustainable Living", the arguments for conserving biodiveristy were summarized thus:

1. As a matter of principle
2. As a matter of survival (i.e., human life support)
3. As a matter of economic benefit (WCU/UNEP/WWF 1991).

Biodiversity can be said to have intrinsic value in its own right, or heritage and amenity value for humans, and can be subject to consumptive or non-consumptive use. There have been numerous discussions in the literature as to which arguments in support of biodiversity should predominate (Livingstone 1981; Norton 1986, 1987; Ehrenfeld 1988; Fox 1990; Randall 1991; Lawton 1991).

Economic arguments basically treat species as human commodities which have some use, and hence economic benefit. This may be already exploited or not yet discovered, in which case species are seen as having "opportunity value". If a species is lost human society has lost the opportunity to exploit it in the future, were some use found for it. While economic arguments are valid for a few important species, they have been criticized as not pertaining to the vast majority of species and of emphasizing exploitation rather than conservation (e.g., Livingstone 1981; Lawton 1991).

Arguments on principles centre on whether humans have the right to utilize and destroy species and whether species have a right to exist, or have intrinsic worth, beyond any potential economic value to humans. Also involved here are aesthetic and quality of life arguments, such as the argument that the diversity of species is beautiful and interesting (Lawton 1991), and compassionate arguments (Ehrlich and Mooney 1983; Ehrlich and Ehrlich 1981). While several authors view ethical and aesthetic arguments as the most important in favour of the conservation of biodiversity (e.g., Ehrenfeld 1988; Norton 1987), others have noted that these arguments have, as yet, had little impact in the real world (Livingstone 1981). Indeed, a complete shift in human attitudes is required for these arguments to become important. Although this is happening to some extent, Soulé (1991) points out that human-oriented, or anthropocentric, values still predominate.

The life support systems argument holds that species diversity is an important component of natural ecosystems and allows them to continue providing ecosystem services to humanity (Ehrlich and Ehrlich 1981), not to mention functioning in their own right. The types of ecosystem services involved include climate control, maintenance of atmospheric composition, freshwater supplies and soils, waste disposal and pest and disease control (Ehrlich and Ehrlich 1981), and as such can be viewed as indirect economic benefits. Ehrlich and Ehrlich (1981) further postulate that, while the loss of one or a few species will probably not affect these functions, the cumulative effect of multiple species losses could be disastrous — their "rivet-popper" analogy (in which an airline employee removes rivets from an airliner's wings on the basis that the plane is manufactured with more rivets than are necessary to fly safely and nothing untoward has happened yet despite his continued removal of rivets). Lawton (1991) criticizes this argument saying that it is likely that maintenance of species diversity is probably not that important for maintaining life support systems and that functioning systems can probably continue with rather few species. Alarm about extinction spasms also needs to be tempered with a consideration of just how few species have gone extinct in recent times (rather than the usual consideration of how many). In Australia, we know of about 100 plants that are presumed extinct, out of a known total of 16 000+, 18 mammals out of a total of 198, one reptile out of about 700 and three birds out of about 720 (Kennedy 1990). Given the degree of transformation that has occurred since European settlement of Australia (Hobbs and Hopkins 1990), the proportion of the total species present which have gone extinct is trivial (although we do not know what proportion of the invertebrates have gone extinct, and the numbers of species threatened with extinction are considerably greater). While this does not in any way reduce the urgency of attending to the current predicament of many species and ecosystems, discussions of the potential importance of loss of biodiversity have to be set within the reality of the situation. It is the argument concerning the importance of species diversity in ecosystem functioning that this book addresses in detail.

Ecosystem functioning

The term ecosystem functioning refers to the capacity of an ecosystem to carry out the primary ecosystem processes of capturing, storing and transferring energy, carbon dioxide, nutrients and water (Solbrig 1991; Woodward 1993). It is a somewhat woolly concept, partially because the concept of ecosystems is itself relatively poorly defined (Burns 1992; Haila *et al.* 1993; see also Schultze and Zwölfer 1987; Schultze and Mooney 1993). Imbedded in these overall ecosystem processes are many more population and community level processes which ensure the continued functioning of the ecosystem (May 1989; O'Neill 1989). Woodward (1993) refers to these as "endogenous properties accruing from functions". Majer (1992) has listed a series of processes or functions central to ecosystem functioning (Table 2). Thus for

Table 2. Principal functions or processes essential for the maintenance of transfers of energy, nutrients and water (modified from Majer 1992).

Function/process	Action
Soil formation	Nutrient cycling
Decomposition	Nutrient cycling
Mycorrhizal activity	Enhanced nutrient uptake
Photosynthesis	Energy capture, water cycling Plant growth, structure, habitat
Herbivory	Energy, nutrient transfer Prevention of dominance
Pollination	Plant reproduction
Propagule dispersal	Plant movement
Predation/parasitism	Energy/nutrient transfer Enhancement of animal diversity
Other species interactions (mutualists/symbionts/competitors)	Energy/nutrient transfer Plant and animal diversity
Response to disturbance	Nutrient transfer. Plant establishment. Prevention of dominance
Hydrologic processes	Water cycling

instance nutrient cycling within an ecosystem may depend on the response of the plant community to disturbance, grazing by herbivores, nitrogen fixing symbioses, plant/pollinator interactions and so on. Thus, while we refer to ecosystem processes in this book, much of the discussion will centre on the subcomponents of the ecosystem since it appears likely that it is at this level that biodiversity is likely to be important (see Hobbs, Chapter 11).

When considering the ecosystem function of biodiversity, it is useful to consider whether species can be grouped according to their ecological function. Solbrig (1991) subdivides species in an ecosystem into different functional types or groups which contain functionally similar taxa or functional analogues. The question then becomes on what basis to classify functional types. There are a variety of ways that species can be grouped on the basis of various aspects of ecosystem function and/or adaptive responses to environmental variables e.g.:

1. Resource use: e.g., litter, seeds, foliage etc.
2. Ecosystem function: production, consumption, decomposition, N fixation
3. Response to disturbance
4. Response to environment: mesophytes, xerophytes etc.
5. Reproductive strategy, pollination systems
6. Tolerance to stress: halophytes etc.

7. Physiognomic types (Raunkier etc.)
8. Physiological types: C_3, C_4 etc.
9. Phenology

It has yet to be seen whether sensible functional groupings can be produced which incorporate some or all of these various classifications. One way to treat the problem is to view the different classifications in a hierarchical perspective (Fig. 3), while multivariate approaches may indicate the existence of groupings arising from the combination of various classifications. Multivariate approaches can themselves be used to provide a hierarchical classification, especially in the case of methods such as Canonical Correspondence Analysis (Ter Braak 1987) which explicitly relate patterns of variation to explanatory variables. Grouping will also largely reflect the purpose of the classification (Körner 1993; Solbrig 1993). The best and most practical approach may, however, be to group species in relation to limiting or dominant ecosystem processes (Walker 1992).

The grouping of species into functional types raises the issue of functional redundancy or substitutability, i.e., the extent to which species within functional groups may be interchangeable and hence to which species diversity is higher than strictly required for ecosystem functioning (di Castri 1991; Solbrig 1991; Walker

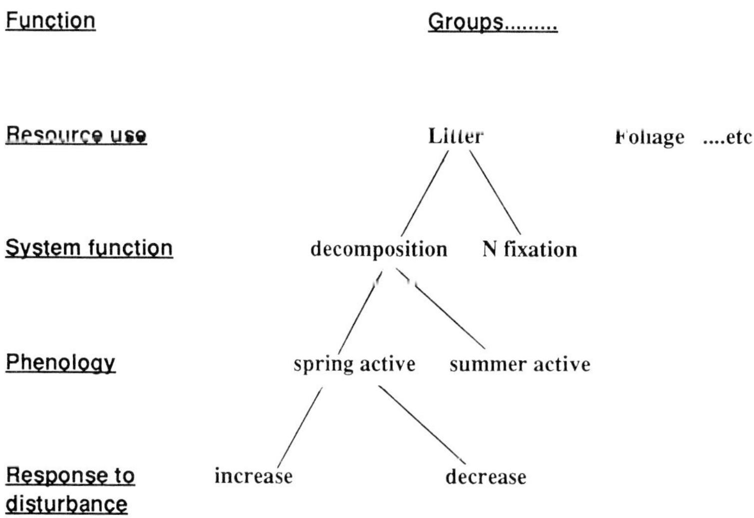

Fig. 3. An example of a hierarchial treatment of ecosystem functional groupings.

1992; Lawton and Brown 1993). Walker (1992) considers that a knowledge of ecological redundancy is essential for the proper evaluation of a decline in biodiversity, and that we need to ascertain which aspects of diversity and which kinds of species are most important to ecosytem function. In particular, identification of functional groups containing little redundancy may aid in directing conservation priorities. Further, it is likely that some species have no functional analogues and are hence "keystone" species (Bond 1993; Lamont, Chapter 6), and it is important to identify these species.

It is possible that redundancy is important for reducing the probability of total system failure, and in fact Main (Chapter 5) argues that provision of back-ups in case of temporary or permanent species loss may be the major function of species diversity in some systems. Ascertaining the levels of redundancy in any given ecosystem is liable to be difficult, since species which are apparently functionally identical may function differently under differing environmental conditions. Hence, a long time perspective will be required to assess redundancy.

Mediterranean Ecosystems in Australia

Physical characteristics

The mediterranean-type climate is characterized by warm/hot summers and cool winters with a more or less well defined summer drought. The extent of mediterranean-type climates in Australia depends on the definition used. Specht (1981), following Dick (1975), uses Köppen climatic zones Csa and Csb to define the areas classed as mediterranean type, but also adds that allied ecosystem types extend into the semi-arid region, classed as BSk (Fig. 4). The distinction between Csa and Csb is in summer temperatures, that in Csa being over 22°C, while Csb is under 22°C. Here we restrict our discussions mostly to the true mediterranean-type regions indicated as Csa and Csb in Figure 4. The main areas are thus the southwestern section of Western Australia and sections near the coast of South Australia and western Victoria. For the purposes of this book, we will concentrate mostly on southwestern Australia. Details of mediterranean ecosystems in eastern Australia can be found in Noble and Bradstock (1989).

Di Castri (1981) gives a more detailed classification of mediterranean-type climates which are split on the basis of the length of summer drought. These are:

1. perarid (11–12 months drought)
2. arid (9–10 months)

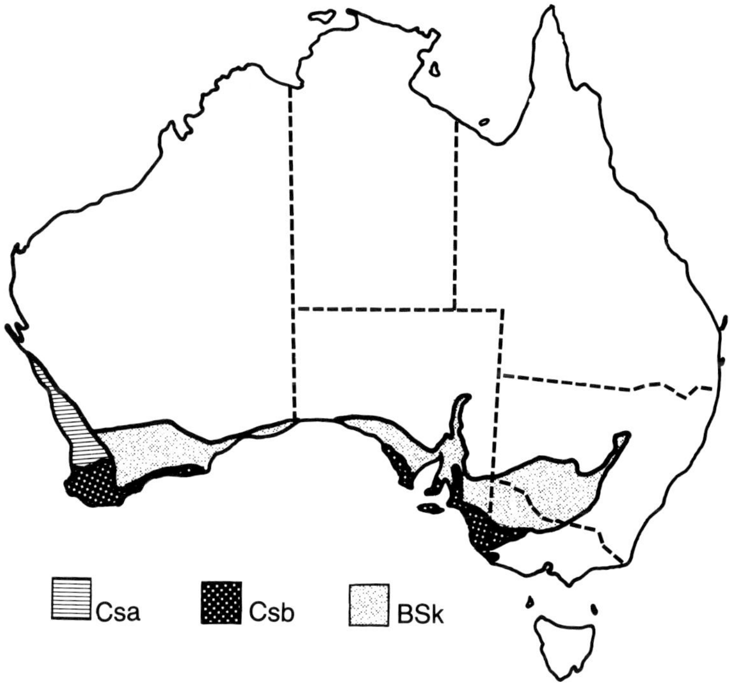

Fig. 4. Köppen climatic zones which define the limits of mediterranean-type climates (Csa and Csb) in southern Australia, with adjoining semi-arid regions (BSk) (after Dick 1975).

3. semi-arid (7–8 months)
4. subhumid (5–6 months)
5. humid (3–4 months)
6. perhumid (1–2 months).

Clearly, these types merge into the arid types at one extreme and the humid types at the other. As such, the mediterranean climate is transitional between temperate and dry tropical climates. Di Castri's classification has some value since strong rainfall gradients are experienced over the mediterranean regions of Australia. In the southwestern region, rainfall declines from about 1 200 mm per year in the extreme south-west (Gentilli 1989) to about 250 mm at the eastern edge of the area which could be classed as mediterranean, and the length of the summer drought increases correspondingly (Fig. 5).

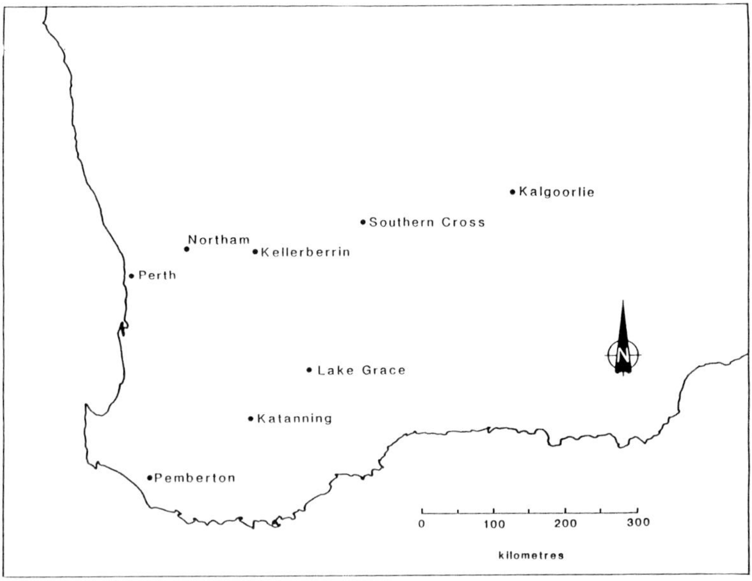
Kalgoorlie
Southern Cross
Northam
Kellerberrin
Perth
Lake Grace
Katanning
Pemberton
N
0 100 200 300
kilometres

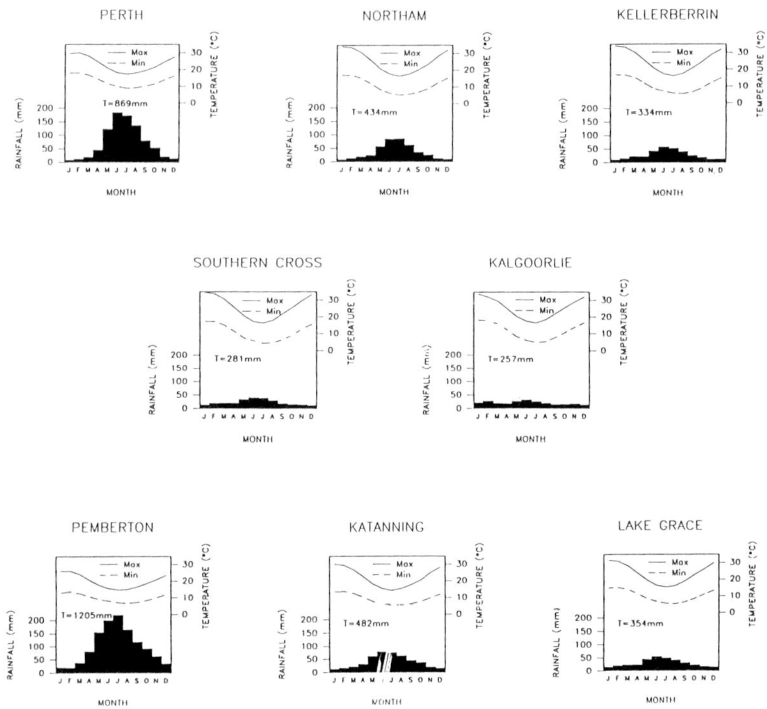
PERTH
T=869mm
NORTHAM
T=434mm
KELLERBERRIN
T=334mm
SOUTHERN CROSS
T=281mm
KALGOORLIE
T=257mm
PEMBERTON
T=1205mm
KATANNING
T=482mm
LAKE GRACE
T=354mm
Max
Min
RAINFALL (mm)
TEMPERATURE (°C)
J F M A M J J A S O N D
MONTH

Di Castri (1981) also points out that Australian mediterranean climates, along with those of South Africa, differ from those of the other mediterranean regions (i.e., the true Mediterranean, California and Chile) because summer rainfall is a common, if unpredictable, occurrence in these regions. The soils of southwestern Australia and South Africa are very old, weathered and nutrient poor, and this again sets them apart from the other regions (di Castri 1981). In common with the other regions, however, is the importance of fire as a factor influencing natural ecosystems (Hunter 1982; Fox and Fox 1986)

Biotic characteristics

Because of the range of rainfalls covered by the mediterranean climate, coupled with variations in soil and landform, the mediterranean area of southwestern Australia encompasses a wide array of ecosystem types, ranging from closed forests to open savannah woodlands and shrublands (Beard 1983, 1990; Fig. 6). The area is characterised by a high level of plant species diversity, both within individual communities and in terms of species turnover (Fig. 7). High levels of diversity are thought to be a result of a number of factors (Hopper 1979; Chapter 2), in particular:

1. A relatively long period of environmental stability and lack of glaciation-related changes (compared with the northen hemisphere mediterranean areas).
2. Isolation from the rest of Australia by the arid Nullarbor Plain.
3. The meeting of neotropical and temperate flora assemblages.
4. The development of a complex mosaic of landforms and soils during the Tertiary and Quaternary, coupled with labile population structures.
5. Interacting disturbance factors, in particular fire and drought, coupled with resource (i.e., nutrient) scarcity (see Hnatiuk and Hopkins 1980; Griffin *et al.* 1983)

High levels of endemism and large numbers of plant species with restricted ranges mean that southwestern Australia contains the highest numbers of rare, endangered and restricted plant species in Australia (Leigh *et al.* 1984; Hopper *et al.* 1990; Hopper, this volume). This therefore makes it a valuable model for the examination of the functioning of biodiversity (see also Pate and

Fig. 5 (opposite). Transects from the coast of Western Australia inland, indicating the changes in rainfall amounts and distributions.

Fig. 6. Examples of vegetation types in mediterranean southwestern Australia. (**a**) Species-rich sandplain heathland, or kwongan. (**b**) Open woodland dominated by *Eucalyptus wandoo*/*E. capillosa* and *E. salmonophloia.* Photographs by R. J. Hobbs.

Fig. 7. Examples of the large array of flowering plants to be found in the mediterranean region of southwestern Australia. (**a**) *Grevillea eriostachya* (**b**) *Hakea coriacea* (**c**) *Beaufortia bracteosa* (**d**) *Eucalyptus caesia* (**e**) *Actinotus superbus*. Photographs by R. J. Hobbs.

Hopper 1993). While the area also retains elements of the mammal fauna which are now missing from the rest of the continent (e.g., Kennedy 1990), levels of faunal diversity and endemism do not, in general, match the high floristic diversity (see Lambeck, Chapter 7).

An important factor to be noted concerning the Australian fauna is the absence of large ungulates. The extinction of the megafauna in the late Quaternary left Australia (in general, not just the mediterranean areas) without large herbivores and predators (Merriles 1984; Murray 1984). This meant that herbivory was carried out primarily by smaller animals and that fire became the major recycler of plant material.

Human transformation

The mediterranean regions of the world differ greatly in their histories of human occupation. Indigenous agriculture and animal husbandry has been practised in the Mediterranean basin for 8–10 000 years, whereas the other regions were occupied only by indigenous hunter-gatherers until relatively recently. European settlement occurred at different times: Chile was colonized in 1550, South Africa in 1650, California in 1769 and southwestern Australia not until 1827 (Fox and Fox 1986). Introduction of European grains and livestock followed thereafter; in Chile this was 430 years ago, South Africa 330 years ago, Australia 150 years ago and California 130 years ago (Fox and Fox 1986). The

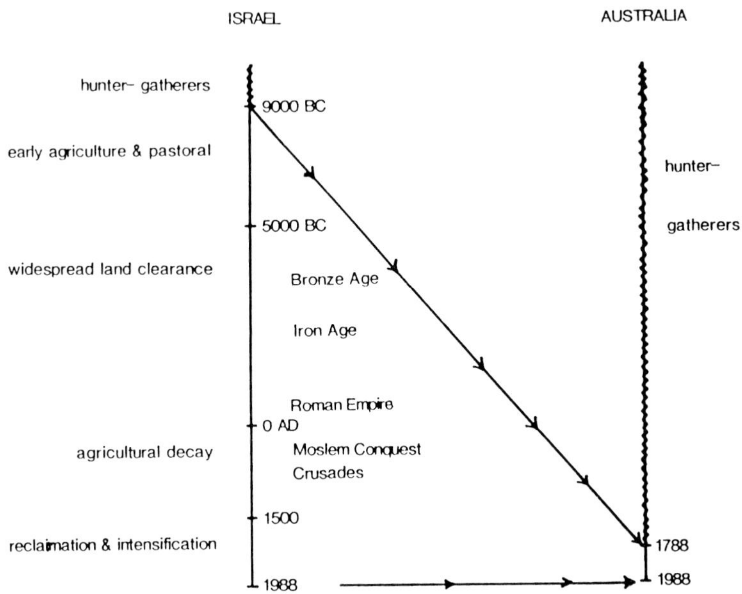

Fig. 8. Modification of natural vegetation has taken place much more rapidly in Australia than in the Mediterranean basin, where present-day patterns of vegetation have developed over thousands of years under the influence of European civilization and in parallel with a gradual intensification of land-use patterns (from Hobbs and Hopkins 1990).

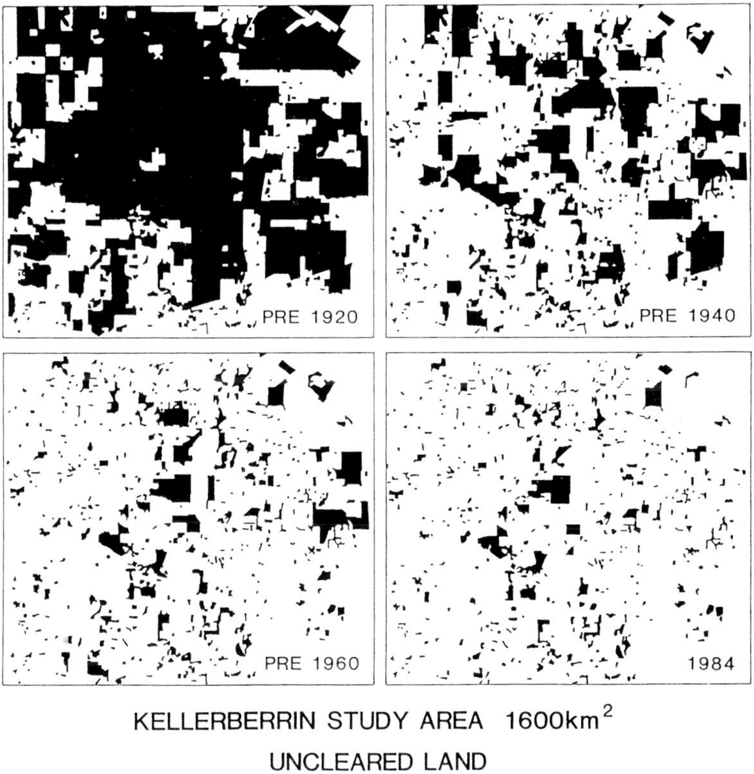

Fig. 9. The development of a 40 × 40 km area of the central wheatbelt of Western Australia for agriculture over the period pre-1920 to 1984. Black areas are natural vegetation, white areas are agricultural land (after Hobbs *et al.* 1992).

duration of settlement has thus been considerably shorter in Australia than in the other regions, although the impact has been profound and rapid (Aschmann 1973; Fox 1982). The transformation of much of the Australian mediterranean-climate areas from predominantly natural vegetation to predominantly agricultural has occurred in a little over a century, compared with 8–10 000 years in the Mediterranean basin (Hobbs and Hopkins 1990; Fig. 8). The rapid transformation has resulted in a highly fragmented landscape, with many small remnants of native vegetation embedded in a predominantly agricultural matrix (Hobbs *et al.* 1992; Saunders and Hobbs, Chapter 4; Majer, Chapter 9; Figs 9, 10).

This level of transformation, coupled with the introduction of domestic and feral animals (e.g., Fig. 11) and exotic plant species, has had dramatic effects on the biota (Hobbs *et al.* 1992), both

Fig. 10. Development for agriculture has left vegetation present mostly in small remnant areas, often completely isolated from other remnants (**a**), or as individual trees in paddocks and along road verges (**b**). Photographs by R. J. Hobbs.

in terms of loss of habitat and species extinctions. The rapidity of change has meant that the biota has not yet responded fully. In the Mediterranean basin, the long presence of technological man has resulted in biotic assemblages which are resilient to human disturbance since they have lost most of their non-resilient components (Fox and Fox 1986). In Australia, however, the rapidity of development has meant that the ecosystems have not yet fully responded to human interference and have retained some of their less resilient components (Hobbs and Hopkins 1990; Fig. 12). We thus have many small remnant areas with biotic assemblages which will probably not be able to persist in the long term without rapid and large-scale efforts to ameliorate the effects of habitat reduction, isolation and changes in basic ecosystem processes (Hobbs and Saunders 1992; Hobbs *et al.* 1992). An important consideration in the context of this book is, however, that the ecosystems of the Mediterranean basin still "function". Does the potential loss of non-resilient species matter, therefore, from the perspective of ecosytem function?

The mediterranean-climate region of southwestern Australia also contains a large forest area, dominated by jarrah *(Eucalyptus marginata)* which has not been subjected to rapid fragmentation, but which has, nevertheless, undergone dramatic modification since European settlement due to unsustainable logging of old growth, modification of fire regimes, mining activities, and the introduction of the pathogen *Phytopthora cinnamomi* (Bartle and Slessar 1989; Bell *et al.* 1989; Dell and Malajczuk 1989; Havel 1989; Havel *et. al* 1989). Mining activities also occur in some of the remaining extensive areas of non-forest vegetation, and *Phytopthora cinnamomi* is now an important problem affecting many other plant communities and threatening to become a major agent of extinction (Wills 1992).

The mediterranean-climate region of southwestern Australia is thus an area of high biotic diversity which has undergone rapid transformation through the human activities of land clearance, forestry, mining and introduction of exotic organisms. These activities continue to have important effects on the biota, which are likely to persist in the future. The effects of land use changes, in themselves, are huge, but if the possible effects of global climate change are also factored in, more potentially large and uncertain

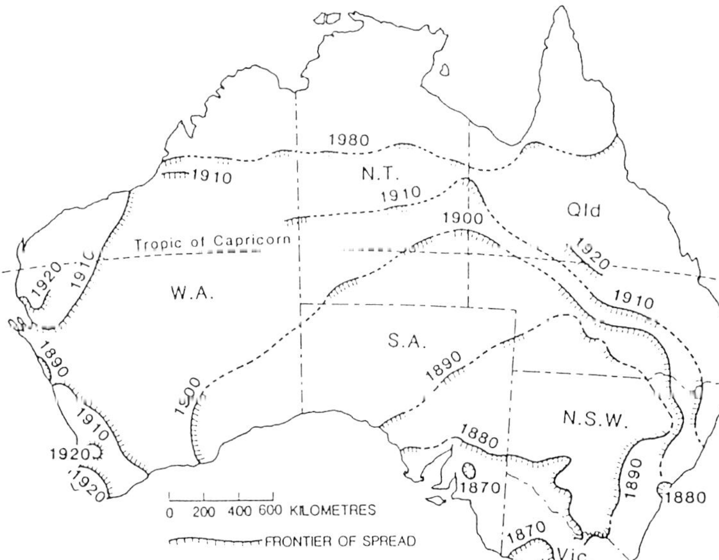

Fig. 11. Spread of the rabbit across Australia following its introduction in south-eastern Australia (after Stodart and Parer 1988).

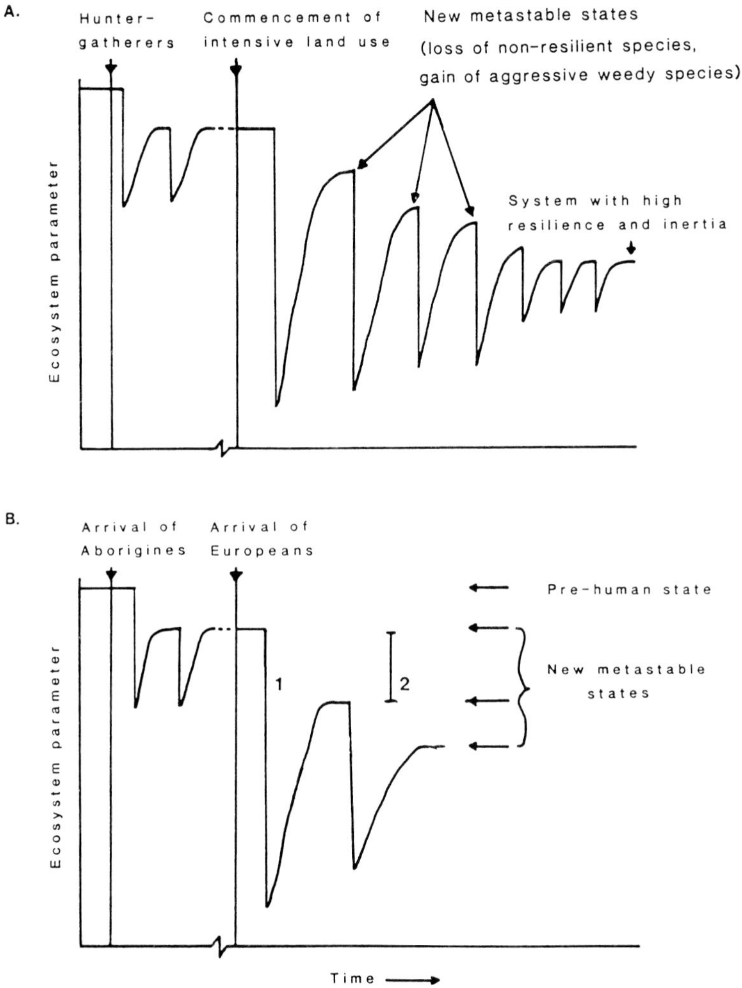

Fig. 12. Resilience and inertia of European ecosystems with a long history of intensive human use versus Australian ecosystems with relatively recent intensive disturbance. In European systems (A), some changes resulted from the development of hunter-gatherer societies, but development of agriculture and more intensive land-use led to greater change and the development of metastable, simplified systems which are now highly resilient and resistent to further change. Australian ecosystems (B) can be viewed in general as being in the early stages of this progression and large changes in vegetation now observed may be the result of response to European disturbance (1) coupled with a permanent change in ecosystem properties (2) (from Hobbs and Hopkins 1990).

changes can be expected. This book therefore examines Australian mediterranean ecosystems in the light of this rapid change, both past and present.

This preliminary investigation into the ecosystem function of biodiversity is thus carried out in the context of extensive human disturbances and ecosystem fragmentation. In this respect the mediterranean-type ecosystems of southwestern Australia represent a paradigm for many areas of the modern world. No longer can we restrict our studies to "pristine" environments. Conservation lands are now imbedded in a production-oriented matrix, and their long-term survival depends on the development of sustainable land-uses and an understanding of potential future responses to land-use and climate change. Biodiversity must thus be set in this context, as recognized in the recent "Sustainable Biosphere Initiative" (Lubchenco *et al.* 1991). This book attempts to do that.

Acknowledgements

I thank Sandra Lavorel, Ted Lefroy, Bert Main, Denis Saunders and Brian Walker for constructive comments on the draft and the other workshop participants for their input of ideas.

References

Aschmann, H., 1973. Man's impact on several regions with Mediterranean climates. Pp. 363–71 *in* Mediterranean type ecosystems, origin and structure ed by F. di Castri and H. A. Mooney. Springer: New York.

Bartle, J. and Slessar, G. C., 1989. Mining and rehabilitation. Pp. 357–77 *in* The jarrah forest. A complex mediterranean ecosystem ed by B. Dell, J. J. Havel and N. Malajczuk. Kluwer: Dordrecht.

Beard, J. S., 1983. Ecological control of the vegetation of Southwestern Australia: moisture versus nutrients. Pp. 66–73 *in* Mediterranean-type ecosystems: the role of nutrients ed by F. J. Kruger, D. T. Mitchell and J. U. M. Jarvis. Springer: New York.

Beard, J. S., 1990. Plant life of Western Australia. Kangaroo Press: Kenthurst, NSW

Bell, D. T., McCaw, W. L. and Burrows, N. D., 1989. Influence of fire on jarrah forest vegetation. Pp. 203–15 *in* The jarrah forest. A complex mediterranean ecosystem ed by B. Dell, J. J. Havel and N. Malajczuk. Kluwer: Dordrecht.

Bond, W. J., 1989. Describing and conserving biotic diversity. Pp. 2–18 *in* Biotic diversity in Southern Africa: concepts and conservation ed by B. J. Huntley. Oxford University Press: Cape Town.

Bond, W. J., 1993. Keystone species. In press *in* Ecosystem function of biodiversity ed by E.-D. Schultze and H. A. Mooney. Springer–Verlag: Heidelberg.

Burns, T. P., 1992. Ecosystem: a powerful concept and paradigm for ecology. *Bull. Ecol. Soc. Amer.* **73:** 39–43.

Cody, M. L., 1975. Towards a theory of continental species diversities: bird distributions over Mediterranean habitat gradients. Pp. 214–57 *in* Ecology and evolution of communities ed by M. L. Cody and J. Diamond. Belknap: Cambridge, Mass.

Dell, B. and Malajczuk, N., 1989. Jarrah dieback — a disease caused by *Phytopthora cinnamomi*. Pp. 67–87 *in* The jarrah forest. A complex mediterranean ecosystem ed by B. Dell, J. J. Havel and N. Malajczuk. Kluwer: Dordrecht.

di Castri, F., 1981. Mediterranean-type shrublands of the world. Pp. 1–52 *in* Ecosystems of the world 11. Mediterranean-type shrublands ed by F. di Castri, D. W. Goodall and R. L. Specht. Elsevier: Amsterdam.

di Castri, F., 1991. Ecosystem evolution and global change. Pp. 189–218 *in* Perspectives in biological complexity ed by O. T. Solbrig and G. Nicolis. IUBS: Paris.

Dick, R. S., 1975. A map of the climates of Australia according to Köppen's principles of definition. *Qld Geogr. J.* **3**: 33–69.

Ehrenfeld, D., 1988. Why put a value on biodiversity? Pp. 212–16 *in* Biodiversity ed by E. O. Wilson. National Academy Press: Washington, DC.

Ehrlich, P. R. and Ehrlich A. H., 1981. Extinction. The causes and consequences of the disappearance of species. Ballantine: New York.

Ehrlich, P. R. and Mooney, H. A., 1983. Extinction, substitution and ecosystem services. *BioScience* **33**: 248–54.

Fox, M. D., 1982. Vegetation changes in Mediterranean Australia since European settlement. Pp. 112–17 *in* Dynamics and management of mediterranean-type ecosystems ed by C. Conrad and W. C. Oechel. *USDA For. Serv. Gen. Tech. Rep. PSW–58.* Pacific Southwest Forest and Range Experiment Station: Berkeley, California.

Fox, B. J. and Fox, M. D., 1986. Resilience of animal and plant communities to human disturbance. Pp. 39–64 *in* Resilience in mediterranean-type ecosystems ed by B. Dell, A. J. M. Hopkins and B. B. Lamont. Junk: Dordrecht.

Fox, W., 1990. Toward a transpersonal ecology. Developing new foundations for environmentalism. Shambala: Boston, Mass.

Gentilli, J., 1989. Climate of the jarrah forest. Pp. 23–40 *in* The jarrah forest. A complex mediterranean ecosystem ed by B. Dell, J. J. Havel and N. Malajczuk. Kluwer: Dordrecht.

Griffin, E. A., Hopkins, A. J. M. and Hnatiuk, R. J., 1983. Regional variation in mediterranean-type shrublands near Enneabba, southwestern Australia. *Vegetatio* **52**: 103–27.

Haila, Y., Saunders, D. A. and Hobbs, R. J., 1993. What do we presently understand about ecosystem fragmentation? *In* Nature Conservation 3: Reconstruction of fragmented ecosystems; global and regional perspectives ed by D. A. Saunders, R. J. Hobbs and P. R. Ehrlich. Surrey Beatty & Sons: Chipping Norton, NSW.

Havel, J. J., 1989. Land use conflicts and the emergence of multiple use. Pp. 281–314 *in* The jarrah forest. A complex mediterranean ecosystem ed by B. Dell, J. J. Havel and N. Malajczuk. Kluwer: Dordrecht.

Havel, J. J., Dell, B. and Malajczuk, N., 1989. Concluding remarks. Pp. 401–03 *in* The jarrah forest. A complex mediterranean ecosystem ed by B. Dell, J. J. Havel and N. Malajczuk. Kluwer: Dordrecht.

Hnatiuk, R. J. and Hopkins, A. J. M., 1980. Western Australian species-rich kwongan (sclerophyllous shrubland) affected by drought. *Aust. J. Bot.* **28:** 573–85.

Hobbs, R. J. and Hopkins, A. J. M., 1990. From frontier to fragments: European impact on Australia's vegetation. *Proc. Ecol. Soc. Aust.* **16:** 93–114.

Hobbs, R. J. and Saunders, D. A., 1992. Conclusions. Can we reintegrate fragmented landscapes? Pp. 299–309 *in* Reintegrating fragmented landscapes. Towards sustainable production and nature conservation ed by R. J. Hobbs and D. A. Saunders. Springer: New York.

Hobbs, R. J., Saunders, D. A. and Arnold, G. W., 1993. Integrated landscape ecology: A Western Australian perspective. *Biol. Conserv.* In press.

Hobbs, R. J., Saunders, D. A., Lobry de Bruyn, L. A. and Main, A. R., 1992. Changes in biota. Pp. 65–106 *in* Reintegrating fragmented landscapes. Towards sustainable production and nature conservation ed by R. J. Hobbs and D. A. Saunders. Springer: New York.

Hopper, S. D., 1979. Biogeographical aspects of speciation in the southwest Australian flora. *Ann. Rev. Ecol. Syst.* **10:** 399–420.

Hopper, S. D., van Leeuwen, S., Brown, A. P. and Patrick, S. J., 1990. Western Australia's endangered flora and other plants for consideration for declaration. Department of Conservation and Land Management: Perth.

Hunter, S. C., 1982. Perspectives of managing mediterranean-type ecosystems: a summary and synthesis. Pp. 64–68 *in* Dynamics and managment of mediterranean-type ecosystems ed by C. E. Conrad and W. C. Oechel. *USDA For. Serv. Gen. Tech. Rep. PSW–58.* Pacific Southwest Forest and Range Experiment Station: Berkeley, California.

Kennedy, M., 1990. Australia's endangered species. The extinction dilemma. Simon and Shuster: Brookvale, NSW.

Körner, C., 1993. Scaling from species to vegetation: the usefulness of functional groups. In press *in* Ecosystem function of biodiversity ed by E.-D. Schultze and H. A. Mooney. Springer–Verlag: Heidelberg.

Lawton, J., 1991. Are species useful? *Oikos* **62:** 3–4.

Lawton, J. H. and Brown, V. K., 1993. Redundancy in ecosystems. In press *in* Ecosystem function of biodiversity ed by E.-D. Schultze and H. A. Mooney. Springer–Verlag: Heidelberg.

Leigh, J. H., Boden, R. and Briggs, J., 1984. Extinct and endangered plants of Australia. MacMillan: Melbourne.

Livingstone, J. A., 1981. The fallacy of wildlife conservation. McClelland and Stewart: Toronto.

Lubchenco, J., Olson, A. M., Brubacker, L. B., Carpenter, S. R., Holland, M. M., Hubbell, S. P., Levin, S. A., MacMahon, J. A., Matson, P. A., Melillo, J. M., Mooney, H. A., Peterson, C. H., Puliam, H. R., Real, L. A., Regal, P. J. and Risser, P. G., 1991. The sustainable biosphere initiative: an ecological research agenda. *Ecology* **72:** 371–412.

Majer, J. D., 1992. Re-establishment of ecosystem functioning in rehabilitated lands. In press *in* Proceedings of Enviromine Conference: Sydney, NSW.

May, R. M., 1989. Levels of organization in ecology. Pp. 339–63 *in* Ecological concepts. The contribution of ecology to an understanding of the natural world ed by J. M. Cherrett, A. D. Bradshaw, F. B. Goldsmith, P. G. Grubb and J. R. Krebs. Blackwell: Oxford.

Mann, C. C., 1991. Extinction: are ecologists crying wolf? *Science* **253:** 736–38.

Merriles, D., 1984. Comings and goings of late Quaternary mammals of extreme southwestern Australia. Pp. 629–38 *in* Quaternary extinctions ed by P. S. Martin and R. G. Klein. University of Arizona Press: Tuscon.

Murray, P., 1984. Extinctions downunder: a bestiary of extinct late Pliestocene monotremes and marsupials. Pp. 600–27 *in* Quaternary extinctions ed by P. S. Martin and R. G. Klein. University of Arizona Press: Tuscon.

Noble, J. C. and Bradstock, R. A. (eds), 1989. Mediterranean landscapes in Australia. Mallee ecosystems and their management. CSIRO: Melbourne.

Norton, B. G. (ed), 1986. The preservation of species. The value of biological diversity. Princeton University Press: Princeton, New Jersey.

Norton, B. G., 1987. Why preserve natural variety? Princeton University Press: Princeton, New Jersey.

Noss, R. F., 1990. Indicators for monitoring biodiversity: a hierarchical approach. *Conserv. Biol.* **4:** 355–64.

O'Neill, R. V., 1989. Perspectives in hierarchy and scale. Pp. 140–56 *in* Perspectives in ecological theory ed by J. Roughgarden, R. M. May and S. A. Levin. Princeton University Press: Princeton, New Jersey.

Pate, J. S. and Hopper, S. D., 1993. Rare and uncommon plants in ecosystems, with special reference to the South West Australian flora. In press *in* Ecosystem function of biodiversity ed by E.-D. Schultze and H. A. Mooney. Springer–Verlag: Heidelberg.

Randall, A., 1991. The value of biodiversity. *Ambio* **20:** 64–68.

Schultze, E.-D. and Mooney, H. A., 1993. The role of biodiversity in ecosystem function. In press *in* Ecosystem function of biodiversity ed by E.-D. Schultze and H. A. Mooney. Springer–Verlag: Heidelberg.

Schultze, E.-D. and Zwölfer, H., 1987. Potentials and limitations of ecosystem analysis. Springer–Verlag: Heidelberg.

Solbrig, O. T., 1991. From genes to ecosystems: a research agenda for biodiversity. IUBS: Paris.

Solbrig, O. T., 1993. Plant traits and adaptive strategies: their role in ecosystem function. In press *in* Ecosystem function of biodiversity ed by E.-D. Schultze and H. A. Mooney. Springer–Verlag: Heidelberg.

Soulé, M. E., 1991. Conservation: tactics for a constant crisis. *Science* **253:** 744–50.

Specht, R. L., 1981. Mallee ecosystems in southern Australia. Pp. 203–31 *in* Ecosystems of the world 11. Mediterranean-type shrublands ed by F. di Castri, D. W. Goodall and R. L. Specht. Elsevier: Amsterdam.

Stodart, E. and Parer, I., 1988. Colonization of Australia by the rabbit *Oryctolagus cuniculatus* (L.). CSIRO, Division of Wildife and Ecology: Canberra.

Taylor, R. W., 1983. Descriptive taxonomy: past, present and future. Pp. 1–47 *in* Australian systematic entomology: a bicentenary perspective. CSIRO: Melbourne.

Ter Braak, C. J. F., 1987. The analysis of vegetation-environment relationships by canonical correspondence analysis. *Vegetatio* **69:** 69–77.

Walker, B. H., 1992. Biodiversity and ecological redundancy. *Conserv. Biol.* **6:** 18–23.

Walker, D., 1989. Diversity and stability. Pp. 115–45 *in* Ecological concepts. The contribution of ecology to an understanding of the natural world ed by J. M. Cherrett, A. D. Bradshaw, F. B. Goldsmith, P. G. Grubb and J. R. Krebs. Blackwell: Oxford.

WCU/UNEP/WWF, 1991. Caring for the earth. A strategy for sustainable living. IUCN: Gland.

Williams, G., 1990. Invertebrate conservation. Pp. 135–43 *in* Australia's endangered species. The extinction dilemma ed by M. Kennedy. Simon and Shuster: Brookvale, NSW.

Wills, R. T., 1992. The ecological impact of *Phytopthora cinnamomi* in the Stirling Range National Park, Western Australia. *Aust. J. Ecol.* in press.

Wilson, E. O. (ed), 1988. Biodiversity. National Academy Press: Washington, DC.

Woodward, F. I., 1993. How many species are required for a functional ecosystem? In press *in* Ecosystem function of biodiversity ed by E.-D. Schultze and H. A. Mooney. Springer–Verlag: Heidelberg.

PLATE 1

ABOVE: Jarrah *(Eucalyptus marginata)* forest. This area has been subjected to fuel reduction burning recently and is also quarantined to prevent spread of the pathogenic fungus *Phytopthora cinammomi.* The forest differs markedly in structure and composition from that present before European exploitation. (Photo: R. J. Hobbs).

BELOW: The grass tree *Xanthorrhoea preisii,* east of York, Western Australia. The above-ground form makes them excellent water channellers (see Groves and Hobbs, Chapter 3). (Photo: R. J. Hobbs).

CHAPTER 2

Patterns of plant diversity at the population and species levels in south-west Australian mediterranean ecosystems

Stephen D. Hopper[1]

Southwestern Australia has remarkable plant species richness (an estimated total of about 8 000), rapid turnover of species across landscapes, and a large number of rare, locally endemic plants. These are the end-products of long and complex evolutionary processes, stimulated by environmental perturbations in the late Tertiary and Quaternary. These perturbations were most pronounced in the Transitional Rainfall Zone, the region where species, infraspecific variants, cryptic species complexes and natural hybrids are most concentrated. The evolution and persistence of so many rare, locally endemic species in the Transitional Rainfall Zone flora suggests that they have a significant role to play in dynamic environments subject to recurrent and unpredictable change. Diversity in ecosystem function is likely to be most pronounced in the plants of the Transitional Rainfall Zone. Caution is urged in applying concepts of population and ecosystem management developed on plants of very recent post-glacial floras from the northern hemisphere to the southwestern Australian flora. Comparatively few southwestern kwongan plants are widespread and easily dispersed over large distances. The overwhelming majority have evolved a myriad of ways to cope with small population sizes, poor dispersal and persistence in a subdued, unglaciated landscape for millions of years. New paradigms are needed for such a remarkably diverse, venerable and poorly documented flora.

Introduction

The flora of southwestern Australia is well known for its richness and high level of endemism. At the species level, an accurate census has yet to be completed, but a figure of around 8 000 seems probable, with about 75 per cent of these found nowhere else.

---

[1]Department of Conservation and Land Management, Western Australian Wildlife Research Centre, PO Box 51, Wanneroo, Western Australia, Australia 6065.
Present Address: Kings Park and Botanic Garden, West Perth, Western Australia, Australia 6005.

Pages 27–46 *in* BIODIVERSITY OF MEDITERRANEAN ECOSYSTEMS IN AUSTRALIA ed by Richard J. Hobbs. Surrey Beatty & Sons, 1992.

Fig. 1. Southwestern Australia showing locations of places mentioned in the text. The dotted line shows the eastern extent of the South-west Botanical Province, after Beard (1980).

The pace of taxonomic and biogeographic discovery has been such that a little under three decades ago, Beard (1965) recorded just 3 611 named taxa for the South-west Botanical Province (stretching in a crescentic arc from Shark Bay to Israelite Bay; Fig. 1). Since then, the number of species in the region recognized in large genera such as *Eucalyptus* and *Acacia* has approximately doubled (though many names have yet to be published), and substantial increases are routine in recent revisionary work across many plant families. For example, George (1991) described 41 new species of feather flowers *(Verticordia)* out of the 97 species he recognized.

Hence, an estimate of 8 000 species in the South-west Botanical Province seems reasonable. This compares favourably with the Cape Floristic Region of South Africa, which has 8 550 species packed into an area about a third the size (Bond and Goldblatt 1984). Both these southern mediterranean-climate regions confound the notion that global plant species richness decreases away from the tropics (Cowling *et al.* 1992).

There have been a number of authoritative reviews on patterns of species richness and diversity in southwestern Australia published over the past decade or so (e.g., George *et al.* 1979; Hopper 1979; Lamont *et al.* 1984; Hopkins and Griffin 1984; Griffin *et al.* 1990; Hopper *et al.* 1992). I intend only to touch briefly on their main conclusions here, and to focus instead on patterns of diversity within species.

My reasons for doing so are twofold. Firstly, there has been a growing realization globally that effective conservation of biodiversity requires information and action at levels of biological organization above and below that of the species (e.g., Falk 1990). Simple-minded notions that once a species is represented in a conservation reserve, it is safe, deny the overwhelming evidence that individual populations are dynamic and part of a larger ever-changing metapopulation. Similarly, individual reserves are components of dynamic landscape systems, where inputs from outside the reserve may override ongoing ecosystem processes within (see Main, Chapter 5; Hobbs, Chapter 11). We must manage metapopulations and whole landscape systems as integrated units to achieve effective conservation of dynamic living organisms.

Secondly, there has been some exciting research completed over recent decades on the population biology and variation within species of the southwestern Australian flora which is yielding new insights of relevance to conservation managers (e.g., James 1982; Moran and Hopper 1987; papers cited in Hopper *et al.* 1990; Pate and Hopper 1993). It is clear that there is far greater diversity in the population genetic architecture, breeding systems and life history strategies of southwestern plants than most conservation managers are aware of. If this diversity is important for the maintenance of ecosystem function in the mediterranean south-west, and I believe it is (Pate and Hopper 1993), an understanding of population level diversity will be crucial for effective conservation.

Phytogeographical context

Beard (1980, 1981) provided a detailed account and maps of phytogeographical districts that have been recognized within the South-west Botanical Province. He advocated the recognition of five such districts, with one covering the higher rainfall forest areas

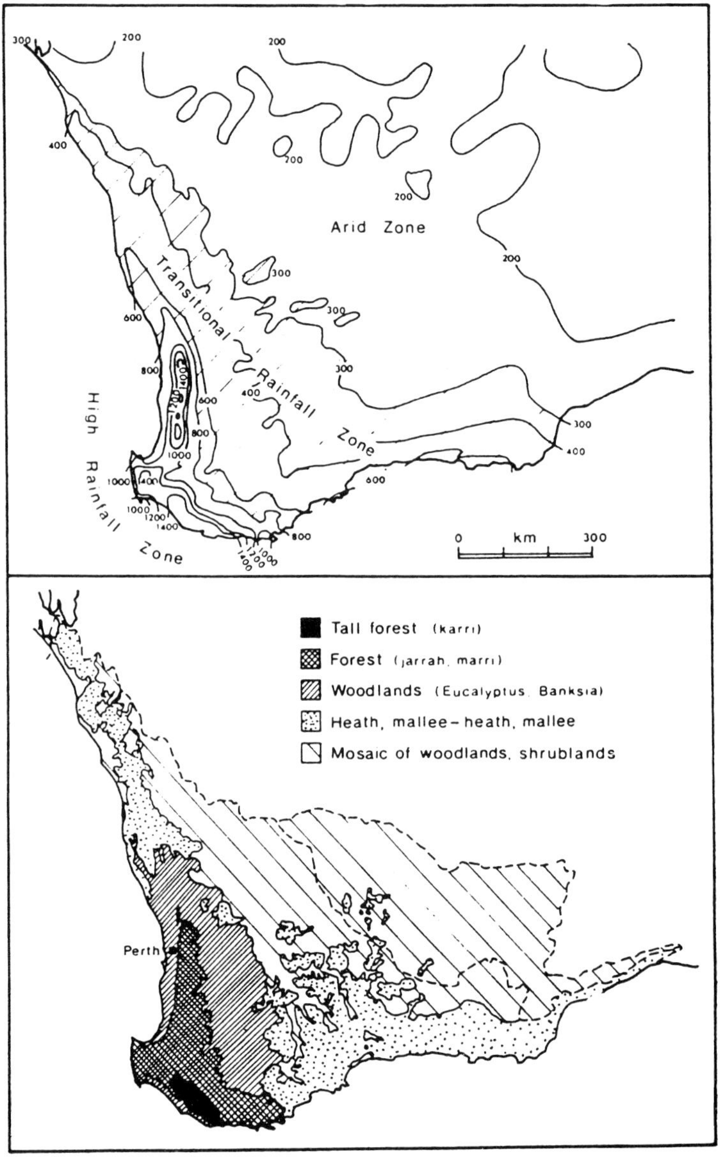

Fig. 2. Distribution of annual rainfall isoheyts (mm), rainfall zones and dominant vegetation formations in southwestern Australia (from Hopper 1979). "Reproduced with permission, from the Annual Review of Ecology and Systematics, Vol. 10, © 1979 by Annual Reviews Inc."

(the Darling District) subdivided into four subdistricts. As yet, this subdivision of the south-west has not been the subject of detailed floristic analysis, although numerical studies of *Eucalyptus* (Gill *et al.* 1985), *Lechenaultia* (Morrison 1987), and *Acacia* (Hnatiuk and Maslin 1988) are a start.

For the purposes of the following discussion, and based on recent biogeographic analysis of granite outcrop floras (Hopper, Ornduff, Brown and Sage, unpubl.), a simpler system which recognizes just three rainfall zones is preferred (Hopper 1979; Fig. 2; see also Hobbs, Chapter 1). These are, the **High Rainfall Zone** (800–1 500 mm annual rainfall), the **Transitional Rainfall Zone** (300–800 mm), and the **Arid Zone** (less than 300 mm). Essentially, the High Rainfall Zone equates to Beard's Darling Botanical District, encompassing forests and woodlands dominated by jarrah *(Eucalyptus marginata),* marri *(E. calophylla),* karri *(E. diversicolor)* and tuart *(E. gomphocephala),* with substantial coastal heath and banksia low woodlands.

The Transitional Rainfall Zone includes the remaining four of Beard's districts in the South-west Botanical Province (the Irwin, Avon, Roe and Eyre), and is the major agricultural wheat-belt of the south-west. It contains the most species-rich areas of wildflowers, including the Mt Lesueur-Kalbarri region to the north of Perth, the lateritic uplands of the western edge of the wheatbelt, and the Stirling Range-Fitzgerald River-Cape Arid National Parks of the southern heathlands and mallee (see Fig. 1 for locations). Vegetation is complex, but dominated by kwongan (shrublands), mallee and woodlands. The Transitional Rainfall Zone also extends inland to encompass some of the Goldfields woodlands and pastoral country beyond the present wheatbelt, including part of Beard's Southwestern Interzone. Further inland again is the Arid Zone, which is largely outside the area of interest of this chapter.

Historical context

A striking feature of the southwestern flora has been its evolution in a landscape which has not been glaciated since the Permian, and in which only modest orogenic activity has occurred subsequently (Hopper 1979, and references therein). The typically flat terrain and subdued landforms of the south-west are the result of millions of years of erosion *in situ.* This has necessitated and allowed, over long periods of time, remarkable adaptive radiation to cope with highly-leached impoverished soils that have been reworked again and again into complex local mosaics. Superimposed on this subtle soil diversity has been major climatic change in the late Tertiary from subtropical to mediterranean, with a concommittant increase in the frequency of fire as a significant ecological process (Martin 1990).

The fossil record indicates that, throughout much of the Tertiary, the south-west was clothed mainly in subtropical rainforests similar to those seen today in Queensland. Fossil deposits (references in Hopper 1979; Hill and Merrifield 1992) as far inland as the Coolgardie area contain material of rainforest gymnosperms in genera such as *Agathis, Araucaria, Dacrydium* and *Podocarpus,* the cycad *Macrozamia,* as well as angiosperm rainforest genera such as *Nothofagus, Alloxylon (Oreocallis), Elaeocarpus* and *Beauprea,* and mangrove genera (*Sonneratia, Avicennia* and *Rhizophora* or *Ceriops*).

Sclerophyllous genera of Myrtaceae and Proteaceae, so dominant today, also formed a component of the Tertiary vegetation, presumably on infertile lateritic soils and in communities marginal to the luxuriant subtropical rainforests (Hill and Merrifield 1992). The lateritic soils were widely established under the seasonal subtropical climate.

Miocene seas laid down the Nullarbor limestones and formed an edaphic barrier to species adapted to acidic soils derived from the granitoid bedrock of the Yilgarn Block. The onset of aridity in the late Tertiary imposed climatic barriers to the north and east, isolating much of the southwestern flora from southeastern congeners, and establishing a mediterranean climate over the south-west with similar zonation to that seen today (Bowler 1982).

Active laterite formation continued only in wetter areas of the High Rainfall Zone. Erosion of lateritic landforms set in elsewhere, leading to differential soil mosaics across the south-west, with greatest local diversity along the western side of the wheatbelt where classic lateritic mesas bounded by breakaways are now archipelago-like in a sea of broad valley slopes and saline valley floors.

Similarly, Quaternary climatic fluctuations had differential spatial impacts, being greatest in the Transitional Rainfall Zone and less pronounced in the High Rainfall Zone and the Arid Zone. The incidence of wildfire undoubtedly increased, and may have been greater in the Transitional Rainfall Zone and Arid Zone than in the mesic High Rainfall Zone.

Sea level fluctuations during the Quaternary were similarly pronounced, but affected near-coastal communities in all three rainfall zones.

In summary, environmental perturbations are likely to have been most frequent and least predictable in the Transitional Rainfall Zone and coastal areas, and less pronounced in the High Rainfall Zone during the late Tertiary and Quaternary. The Arid Zone was predictably harsh throughout this period.

Floristic response and pattern

The present-day flora comprises the end-products of long and complex evolutionary processes stimulated by the major geohistorical circumstances described above. Some striking patterns of differential diversity are evident.

Patterns at the species level

The flora of the south-west is dominated by speciose genera of woody perennials in families such as the Myrtaceae, Proteaceae, Fabaceae, and Epacridaceae. There are few relicts of the Tertiary rainforest flora extant, and only a modest diversity of families and genera remains (Lamont *et al.* 1984). However, the flora has radiated extensively at the species level. In the south-west, *Acacia* has at least 400+ species, *Eucalyptus* 300+, *Grevillea* 200+, *Stylidium* and *Melaleuca* 150+, *Hakea* and *Caladenia* 100+. The serious student of botany has to specialize on selected genera because there are so many species in each and their patterns of variation are so complex. This is in stark contrast to the rainforest flora of Queensland, where there is great diversity in families and genera but relatively few species within genera.

It is clear that explosive speciation has occurred in the south-west in response to the environmental perturbations of the late Tertiary and Quaternary. Moreover, the Transitional Rainfall Zone is where such speciation has been concentrated for the majority of woody perennial taxa (Hopper 1979; Lamont *et al.* 1984).

For almost a century, it has been recognized that there are exceptionally rich floras in the Stirling Range-Fitzgerald River National Park region near the south coast, and in the Mt Lesueur National Park area to the north of Perth. For example Fitzgerald River National Park has 1 750+ named species (Chapman and Newbey 1987), and Mt Lesueur National Park 820+ (Griffin *et al.* 1990). More than 110 species have been found in 10 m × 10 m quadrats near Mt Lesueur, and averages around 80 are common.

Another notable feature is the high turnover of species across the landscape, or gamma diversity. Adjacent quadrats on laterite in Mt Lesueur National Park may have as few as 60 per cent of their species in common. Moving as little as 0.5 km away within the same vegetation type can reduce the similarity to less than 40 per cent (Griffin *et al.* 1990).

At the regional level, there are major differences in the concentration of genera, and recurrent examples of allopatric replacement series of closely related narrowly endemic species within genera. For example, genera such as *Calothamnus, Melaleuca, Hakea, Darwinia* and *Dryandra* show bimodal concentrations of species in the two major nodes mentioned above. Others are concentrated in the southern kwongan and mallee (e.g., *Adenanthos, Eucalyptus, Banksia, Leucopogon*), or the Mt Lesueur-Kalbarri northern kwongan *(Grevillea, Conostylis, Lechenaultia),* or are concentrated inland through the wheatbelt (e.g., *Acacia, Verticordia).*

It is unusual to find species that occur throughout much of the Transitional Rainfall Zone. The majority of known examples have good means of seed dispersal (e.g., the annual composites *Quinettia urvillei, Millotia tenuifolia;* the orchids *Spiculaea ciliata, Caladenia flava, Leporella fimbriata;* the mistletoes *Amyema miquellii, Nuytsia floribunda).*

It is much more common for species to be confined to smaller ranges within the Transitional Rainfall Zone, often in allopatric replacement series with congeners (e.g., the superspecies of *Eucalyptus* series *Levispermae* treated by Brooker and Hopper 1991; Fig. 3). Indeed, the flora has Australia's highest concentration of rare and endangered local endemics, and is exceptional in this respect from a global perspective as well (Hopper *et al.* 1990).

It seems likely that the Transitional Rainfall Zone flora has had to cope with small population sizes induced by recurrent late Tertiary-Quaternary climatic and sea-level oscillations in a heterogeneous soil mosaic. Consequently, the south-west has much to teach us about rare species and their roles in ecosystem function (Pate and Hopper 1993).

Although most of the southwestern flora is concentrated in the Transitional Rainfall Zone, a few genera such as *Agonis* favour the wetter conditions of the High Rainfall Zone. Indeed, a recent account of the flora of the karri forest and adjacent communities

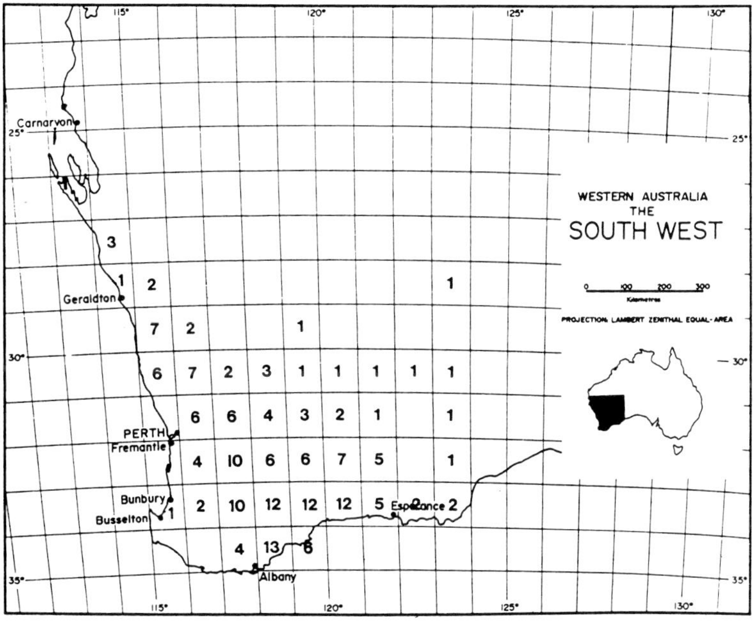

Fig. 3. Distribution and density in 1 × 1 degree squares of taxa of *Eucalyptus* ser. *Levispermae* in southwestern Australia (from Brooker and Hopper 1991).

(Warren Botanical Subdistrict — Hopper *et al.* 1992) listed 1 947 taxa, the most important families being Fabaceae (165 taxa), Orchidaceae (152), Myrtaceae (125) and Proteaceae (115), with the largest genera *Acacia* (64), *Stylidium* (55), *Caladenia* (46), *Leucopogon* (41), *Eucalyptus* (26) and *Drosera* (24). Most groups of wetland monocotyledons, including genera of Cyperaceae, Xyridaceae, Juncaginaceae, Restionaceae and Orchidaceae, are species-rich in the karri region.

It is in this highest rainfall south-coastal area that the few relicts of rainforest affinities are concentrated (e.g., *Cephalotus follicularis, Eucalyptus brevistylis* — see Ladiges *et al.* 1987). Moreover, there are several rare taxa persisting in mesic sites with rainfall more or less year round whose closest affinities may be with eastern Australian summer rainfall taxa (e.g., the banksias described by Hopper (1989)).

Patterns within species

Although studies of variation within species have had a shorter history than the accumulation of knowledge at the species level, a wealth of pattern and process has been documented.

It seems likely that the concentration of species seen in the Transitional Rainfall Zone is a trend replicated infraspecifically in many cases. Morphologically uniform subspecies in the High Rainfall Zone surrounded by allopatric replacement series of subspecies in the adjacent Transitional Rainfall Zone are seen, for example, in *Conostylis aculeata* (Hopper *et al.* 1987), *Anigozanthos bicolor* (Hopper 1987), and *Acacia varia* (Maslin 1975).

Similarly, cryptic speciation involving allopatric replacement series throughout the Transitional Rainfall Zone has been documented in a number of groups. Coates' (1981) study of the *Stylidium caricifolium* complex, and Brooker's (1979, 1988) work on the *Eucalyptus foecunda* group are exemplary. I am unaware of such pronounced replacement series being found in studies of High Rainfall Zone species of plants.

The evolution of new subspecies within the Transitional Rainfall Zone has been documented in the granite outcrop endemics *Eucalyptus caesia* and *E. crucis* (Brooker and Hopper 1982; Hopper *et al.* 1982, 1985; Hopper and Burgman 1983; Moran and Hopper 1983; Sampson *et al.* 1988; Fig. 4). In *E. caesia,* it seems likely that populations nearest the High Rainfall Zone such as at Boyagin Rock stem from the oldest stock, and that there has been progressive migration eastwards into the wheatbelt, culminating in the evolution of the large flowered subspecies *magna.* Similarly in *E. crucis,* the most derived subspecies (ssp. *crucis*) has the more easterly distribution.

As another indicator of enhanced evolutionary activity, natural hybridization has been found to be pronounced in the Transitional Rainfall Zone in genera such as *Eucalyptus* (e.g., Hopper *et al.* 1978; Brooker and Hopper 1991), *Conostylis* (Hopper *et al.* 1987), *Anigozanthos* (Hopper 1977, 1978), and *Caladenia* (Heberle 1982; Hopper and Brown 1993). The evolution of hybrid species through allopolyploidy seems likely in the case of *Conostylis stylidioides* from the Geraldton-Shark Bay area. This species is tetraploid, and resembles closely the diploid natural hybrids of *C. prolifera* and *C. candicans,* both of which grow sympatrically with *C. stylidioides* (Hopper *et al.* 1987).

Chromosomal patterns of infraspecific variation have been found in a few taxa. The penetrating studies of the evolution of translocation heterozygosity in selfing populations of the granite rock herb *Isotoma petraea* provide a well-documented example (James 1965, 1970, 1982, 1984; Beltran and James 1974; James *et al.* 1983, 1989, 1990; Lavery and James 1986). *I. petraea* occurs

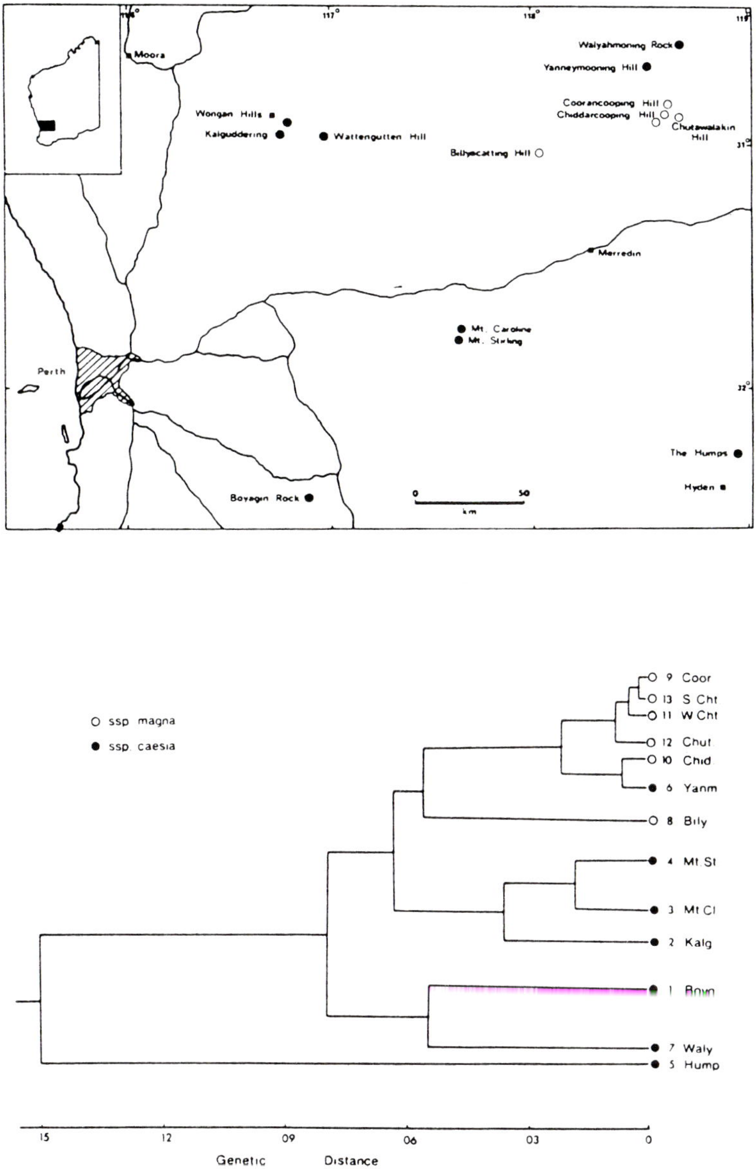

Fig. 4. Distribution of *Eucalyptus caesia* ssp. *caesia* (●) and ssp. *magna* (○) in southwestern Australia, and a dendrogram showing Nei's genetic distance relationship of six populations of ssp. *caesia* and seven of ssp. *magna,* based on allozyme frequencies at 18 loci (from Hopper *et al.* 1982).

throughout much of central Australia in populations that have seven pairs of chromosomes which undergo normal bivalent formation and behaviour at meiosis. At the southwestern end of the species' range, however, as it enters the Transitional Rainfall Zone, translocation heterozygosity has evolved and increasingly larger numbers of the chromosome pairs get tied up in rings at meiosis. All seven pairs are involved in a large ring formation at the extreme southwestern boundary of the range.

This chromosome behaviour is associated with inbreeding and the presence of recessive lethals. James has marshalled a convincing body of evidence that the chromosomal rings conserve heterozygous adaptive genotypes in the face of inbreeding, and have been strongly selected despite more than 90 per cent of the seed being aborted as a consequence of mismatched chromosomal segments in the gametes of the translocation heterozygotes. Allozyme variation in this species is complex, and understandable only in the light of a detailed knowledge of the genetic system of the species (James *et al.* 1983).

Another remarkable case, this time of karyotypic race formation, was documented by Coates and James (1979) for *Stylidium crossocephalum,* a morphologically uniform triggerplant of the kwongan to the north of Perth. Sixteen chromosome races were found between Yanchep and Geraldton (Fig. 5), most of them producing highly sterile hybrids where their ranges overlapped or when crossed in the glasshouse.

Polyploid races and the presence of B chromosomes were found in *Dampiera linearis* by Bousfield (1970; Bousfield and James 1976). This species occurs mainly in the High Rainfall Zone. Similar polyploid races occur in *Conostylis setigera,* another species favouring the wetter forests of the south-west (Hopper *et al.* 1987). These two species constitute some of the few documented cases of genetic differentiation within High Rainfall Zone plants.

Geographical variation within southwestern species has also been documented through allozyme studies. In general, high levels of variability have been detected in these studies. However, some interesting patterns, correlated with life-history attributes, are emerging.

The cycad *Macrozamia riedlei* displayed low levels of interpopulational differentiation but high levels of polymorphism within populations distributed from Walpole to Cataby (Byrne

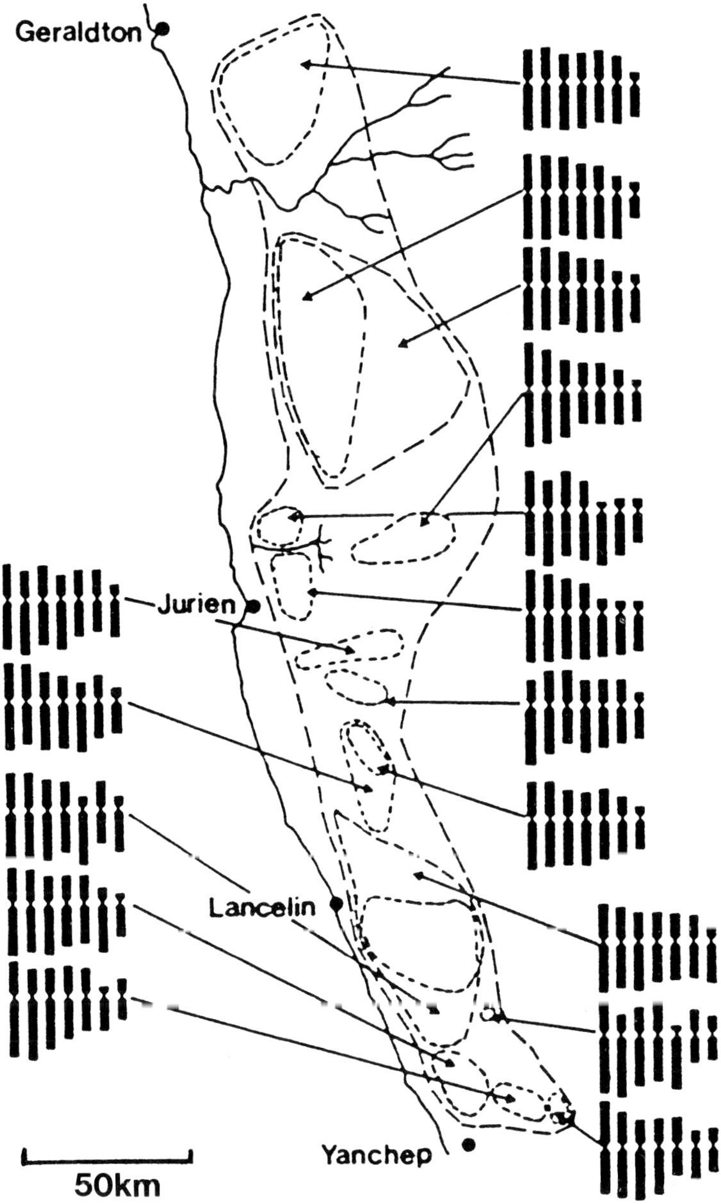

Fig. 5. Distribution of chromosomal variants in the trigger plant *Stylidium crossocephalum* (after Coates and James 1979).

and James 1991). A similar pattern was found in the forest tree *Eucalyptus diversicolor* (Coates and Sokolowski 1989). There was slight differentiation in allozyme frequencies in outlying populations such as at the Porongorups, but most of the detected variation was found within populations.

Coates and Hnatiuk (1990) examined 44 populations of 19 taxa of the woody perennial shrub genus *Eremaea,* which has speciated prolifically in kwongan to the north of Perth. They also found that most allozyme variation was within rather than between populations within species.

Moran and Hopper (1983) and Sampson *et al.* (1988) showed that the converse was true in disjunct populations of the wheatbelt granite rock endemics *Eucalyptus caesia* and *E. crucis* respectively. For these species, variability within populations was relatively low, and differences between populations high, compared with forest eucalypts.

Moran and Hopper (1987) found no clear pattern when they examined allozyme variation in four species of locally endemic eucalypts from the Mt Lesueur area. Rarity and very small population sizes did not necessarily mean depauperate genetic variation, as classical population genetic theory would predict.

Coates (1988) did find some striking differences in the distribution of variation in the endangered subshrub *Acacia anomala,* which is confined to the Darling Scarp in suburban Perth. Populations that were clonal had low variability within and high variability between clumps, whereas populations that were comprised of obligate seeders displayed the converse pattern.

Sampson (1988; Sampson *et al.* 1989) found slight differentiation among disjunct populations of *Eucalyptus lane-poolei* between Serpentine and Badgingarra, and of *E. rhodantha* between Watheroo and Three Springs. However, most variability was again within populations of these two species.

Coates (1992) has recently documented very high levels of interpopulational allozyme divergence in the Wongan Triggerplant *Stylidium coroniforme.* This is associated with founder-flush cycles following disturbance or fire.

Other patterns of geographical variation have been investigated in several southwestern plants. For example, Ornduff (1986) explored the comparative fecundity of heterostylous and non-heterostylous species of *Villarsia* (Menyanthaceae). He found variable levels of seed set among populations of the same species, and no clear increase in fecundity associated with heterostyly.

Walker and Pate (1986) investigated population divergence in leafy versus phyllodinous growth of seedlings and its ecophysiological significance in the widespread shrub legume *Viminaria juncea*. They found that the leafy seedlings outgrew phyllodinous ones in flooded or mesic conditions. This is a useful example where divergence in ecosystem function is occurring within a species. Much more research of this kind is needed, however, before the value of infraspecific variation to the maintenance of ecosystem function in southwestern Australian plants becomes apparent.

Conclusions

What generalizations might be made in relation to ecosystem function about the patterns of diversity discussed above? The remarkable species richness, rapid turnover of species across southwestern landscapes, and large number of rare, locally endemic plants in southwestern Australia are the end-products of long and complex evolutionary processes, stimulated by environmental perturbations in the late Tertiary and Quaternary. These perturbations were most pronounced in the Transitional Rainfall Zone, the region where species, infraspecific variants, cryptic species complexes and natural hybrids are most concentrated.

Detailed study of genetic variation within species has revealed a diversity of patterns, some correlated with life-history attributes or population architecture. In eucalypts, for example, forest species show high allozyme variability within populations and low divergence between populations, whereas species from disjunct wheatbelt granite rocks show higher variability between populations. These patterns conform to classical population genetic expectations. However, the patterns shown by rare eucalypt species are idiosyncratic, and in species such as the granite rock herb *Isotoma petraea*, allozyme patterns depart significantly from expectations due to inbreeding and translocation heterozygosity.

The evolution and persistence of so many rare, locally endemic species in the Transitional Rainfall Zone flora suggests that they have a significant role to play in dynamic environments subject to recurrent and unpredictable change (Main 1982; Pate and Hopper 1993). Assessment of the implications of this geographic and genetic variation for ecosystem functioning is an important undertaking which has, as yet, received little attention. For instance, do allopatric congeners fulfill the same ecosystem functions or are they functionally, as well as genetically, distinct?

I would urge caution in applying concepts of population and ecosystem management developed on plants of very recent post-glacial floras from the northern hemisphere to the southwestern Australian flora. Comparatively few southwestern kwongan plants are widespread and easily dispersed over large distances. The overwhelming majority have evolved a myriad of ways to cope with small population sizes, poor dispersal and persistence in a subdued, unglaciated landscape for millions of years. Such distributions and small population sizes mean that conservation will be dependent on retention of areas outside designated nature reserves. Replication within the reserve system will not always be possible, and management specifically for small, isolated populations is required. If the types of polymorphism described here are general, simple replication of a species in the reserve system may not be sufficient to conserve genetic diversity in any case. New paradigms are needed for such a remarkably diverse, venerable and poorly documented flora. I suspect that we and those who come after us have much to learn about the role of plant biodiversity in ecosystem function from this wonderful heritage.

Acknowledgements

I am grateful to Sid James, Jane Sampson, Dave Coates, Ian Brooker, Andrew Brown and numerous other colleagues for their insight, discussion and collaboration in studies on the southwestern flora, to Denis Saunders and Bert Main for comments on the draft manuscript, and to Richard Hobbs for his remarkable patience.

References

Beard, J. S., 1965. A descriptive catalogue of West Australian plants. Society for Growing Australian Plants: Sydney.

Beard, J. S., 1980. A new phytogeographic map of Western Australia. *West. Aust. Herb. Res. Notes* **3**: 37–58.

Beard, J. S., 1981. Vegetation of Western Australia. 1:1 000 000 series. Swan. University of Western Australia: Nedlands.

Beltran, I. C. and James, S. H., 1974. Complex hybridity in *Isotoma petraea* IV. Heterosis in interpopulational heterozygotes. *Aust. J. Bot.* **22**: 251–64.

Bond, P. and Goldblatt, P., 1984. Plants of the Cape flora. A descriptive catalogue. *J. S. African Bot. Suppl.* **13**: 1–455.

Bousfield, L. R., 1970. Chromosome races in *Dampiera linearis* R.Br. Ph.D. Thesis: University of Western Australia: Nedlands.

Bousfield, L. R. and James, S. H., 1976. The behaviour and possible cyto-evolutionary significance of B chromosomes in *Dampiera linearis* (Angiospermae: Goodeniaceae). *Chromosoma* **55**: 309–23.

Bowler, J. M., 1982. Aridity in the late Tertiary and Quaternary of Australia. Pp. 35–45 *in* Evolution of the flora and fauna of arid Australia ed by W. R. Barker and P. J. M. Greenslade. Peacock Publications: Adelaide.

Brooker, M. I. H., 1979. A revision of the informal series *Foecundae* Pryor and Johnson of the genus *Eucalyptus* L'Herit and notes on variation in the genus. *Brunonia* **2**: 125–70.

Brooker, M. I. H., 1988. *Eucalyptus foecunda* revisited and six related new species (Myrtaceae). *Nuytsia* **6**: 325–34.

Brooker, M. I. H. and Hopper, S. D., 1982. New subspecies in *Eucalyptus caesia* and *E. crucis* (Myrtaceae) of Western Australia. *Nuytsia* **4**: 113–28.

Brooker, M. I. H. and Hopper, S. D., 1991. A taxonomic revision of *Eucalyptus wandoo, E. redunca,* and allied species (*E.* series *Levispermae* — Myrtaceae) in Western Australia. *Nuytsia* **8**: 1–189.

Byrne, M. and James, S. H., 1991. Genetic diversity in the cycad *Macrozamia riedlei*. *Heredity* **67**: 35–39.

Chapman, A. and Newbey, K. R., 1987. A biological survey of the Fitzgerald area, Western Australia. Report to the Australian Heritage Commission: Perth.

Coates, D. J., 1981. Chromosome, morphometric and breeding system studies in the *Stylidium caricifolium* species complex (Stylidiaceae). *Aust. J. Bot.* **29**: 397–417.

Coates, D. J., 1988. Genetic diversity and population genetic structure in the rare Chittering grass wattle *Acacia anomala*. *Aust. J. Bot.* **36**: 273–86.

Coates, D. J., 1992. Genetic consequences of a bottleneck and spatial genetic structure in the triggerplant *Stylidium coroniforme* (Stylidiaceae). *Heredity* **69**: in press.

Coates, D. J. and Hnatiuk, R. J., 1990. Systematic and evolutionary inferences from isozyme studies in the genus *Eremaea* (Myrtaceae). *Aust. Syst. Bot.* **3**: 59–74.

Coates, D. J. and James, S. H., 1979. Chromosome variation in *Stylidium crossocephalum*. F. Muell. (Angiospermae: Stylidiaceae) and the dynamic co-adaptation of its lethal system. *Chromosoma* **72**: 357–76.

Coates, D. J. and Sokolowski, R. E. S., 1989. Genetic diversity in karri *Eucalyptus diversicolor*. *Aust. J. Bot.* **37**: 145–56.

Cowling, R. M., Holmes, P. M. and Rebelo, A. G., 1992. Plant diversity and endemism. Pp. 62–112 *in* The ecology of fynbos: nutrients, fire and diversity ed by R. M. Cowling. Oxford University Press: Cape Town.

Falk, D. A., 1990. Integrated strategies for conserving plant genetic diversity. *Ann. Missouri Bot. Gard.* **77**: 38–47.

George, A. S., 1991. New taxa, combinations and typifications in *Verticordia* (Myrtaceae: Chamelaucieae). *Nuytsia* **7**: 231–394.

George, A. S., Hopkins, A. J. M. and Marchant, N. G., 1979. The heathlands of Western Australia. Pp. 211–30 *in* Heathlands and related shrublands of the world. A. Descriptive studies ed by R. L. Specht. Elsevier: Amsterdam.

Gill, A. M., Belbin, L. and Chippendale, G. M., 1985. Phytogeography of *Eucalyptus* in Australia. Aust. Flora and Fauna Ser. 5. Australian Government Publishing Service: Canberra.

Griffin, E. A., Hopper, S. D. and Hopkins, A. J. M., 1990. Flora. Pp. 39–69 *in* Nature conservation, landscape and recreation values of the Lesueur area ed by A. A. Burbidge, S. D. Hopper and S. van Leeuwen. Environmental Protection Authority: Perth.

Heberle, R., 1982. *Caladenia* in Western Australia and natural hybridization. *The Orchadian* **7**: 78–83.

Hill, R. S. and Merrifield, H. E., 1992. An early tertiary macroflora from West Dale, Western Australia. *Alcheringa,* in press.

Hnatiuk, R. J. and Maslin, B. R., 1988. Phytogeography of *Acacia* in Australia in relation to climate and species-richness. *Aust. J. Bot.* **36**: 361–83.

Hopkins, A. J. M. and Griffin, E. A., 1984. Floristic patterns. Pp. 69–83 *in* Kwongan: plant life of the sandplain ed by J. S. Pate and J. S. Beard. University of Western Australia Press: Nedlands.

Hopper, S. D., 1977. The structure and dynamics of a hybrid population of *Anigozanthos manglesii* D. Don and *A. humilis* Lindl. (Haemodoraceae). *Aust. J. Bot.* **25**: 413–22.

Hopper, S. D., 1978. Speciation in the kangaroo paws of southwestern Australia (*Anigozanthos* and *Macropidia:* Haemodoraceae). Ph.D. Thesis, University of Western Australia: Nedlands.

Hopper, S. D., 1979. Biogeographical aspects of speciation in the south west Australian flora. *Ann. Rev. Ecol. Syst.* **10**: 399–422.

Hopper, S. D., 1987. *Blancoa, Anigozanthos, Macropidia. Flora of Australia* **45**: 110–28.

Hopper, S. D., 1989. New subspecies of *Banksia seminuda* and *B. occidentalis* (Proteaceae) from the south coast of Western Australia. *Nuytsia* **7**: 15–24.

Hopper, S. D. and Brown, A. P., 1993. New and reinstated genera, species, subspecies and hybrids of Western Australian Orchidaceae. *Aust. Orchid Res.*, in press.

Hopper, S. D. and Burgman, M. A., 1983. Cladistic and phenetic analyses of phylogenetic relationships among populations of *Eucalyptus caesia. Aust. J. Bot.* **31**: 35–49.

Hopper, S. D., Campbell, N. A. and Caputi, N., 1985. Geographical variation, subspecies discrimination and evolution in fruits, leaves and buds of *Eucalyptus caesia* (Myrtaceae). *Nuytsia* **5**: 179–94.

Hopper, S. D., Campbell, N. A. and Moran, G. F., 1982. *Eucalyptus caesia,* a rare mallee of granite rocks from southwestern Australia. Pp. 46–61 *in* Species at risk: research in Australia ed by R. H. Groves and W. D. L. Ride. Australian Academy of Science: Canberra.

Hopper, S. D. and Coates, D. J., 1990. Conservation of genetic resources in Australia's flora and fauna. *Proc. Ecol. Soc. Aust.* **16**: 567–77.

Hopper, S. D., Coates, D. J. and Burbidge, A. H., 1978. Natural hybridization and morphometric relationships between three mallee eucalypts in the Fitzgerald River National Park, WA. *Aust. J. Bot.* **26**: 319–33.

Hopper, S. D., Purdie, R. W., George, A. S. and Patrick, S. J., 1987. *Conostylis. Flora of Australia* **45**: 57–110.

Hopper, S. D., van Leeuwen, S., Brown, A. P. and Patrick, S. J., 1990. Western Australia's endangered flora. Department Conservation and Land Management: Perth.

Hopper, S. D., Keighery, G. J. and Wardell-Johnson, G., 1992. Flora of the karri forest and other communities in the Warren Botanical Subdistrict of Western Australia. *West. Aust. Dept. Cons. Land Mgmt. Tech. Rept.*, in press.

James, S. H., 1965. Complex hybridity in *Isotoma petraea* I. The occurrence of interchange heterozygosity, autogamy and a balanced lethal system. *Heredity* **20**: 341–53.

James, S. H., 1970. Complex hybridity in *Isotoma petraea* II. Components and operation of a possible evolutionary mechanism. *Heredity* **25**: 53–77.

James, S. H., 1982. The relevance of genetic systems in *Isotoma petraea* to conservation practice. Pp. 63–71 *in* Species at risk: research in Australia ed by R. H. Groves and W. D. L. Ride. Australian Academy of Science: Canberra.

James, S. H., 1984. The pursuit of hybridity and population divergence in *Isotoma petraea*. Pp. 169–77 *in* Plant biosystematics ed by W. F. Grant. Academic Press: Toronto.

James, S. H. and Hopper, S. D., 1981. Speciation in the Australian flora. Pp. 361–81 *in* The biology of Australian plants ed by J. S. Pate and A. J. McComb. University of Western Australia Press: Nedlands.

James, S. H., Sampson, J. F. and Playford, J., 1989. Complex hybridity in *Isotoma petraea* VII. Assembly of the genetic system in the O6 Pigeon Rock population. *Heredity* **64**: 289–95.

James, S. H., Playford, J. and Sampson, J. F., 1990. Complex hybridity in *Isotoma petraea* VIII. Variation for seed aborting lethal genes in the O6 Pigeon Rock population. *Heredity* **66**: 173–80.

James, S. H., Wylie, A. P., Johnson, M. S., Carstairs, S. A. and Simpson, G. A., 1983. Complex hybridity in *Isotoma petraea* V. Allozyme variation and the pursuit of hybridity. *Heredity* **51**: 653–63.

Ladiges, P. Y., Humphries, C. J. and Brooker, M. I. H., 1987. Cladistic and biogeographic analysis of Western Australian species of *Eucalyptus* L'Herit. Informal Subgenus *Monocalyptus* Pryor and Johnson. *Aust. J. Bot.* **35**: 251–81.

Lamont, B. B., Hopkins, A. J. M. and Hnatiuk, R. J., 1984. The Flora-composition, diversity and origins. Pp. 27–50 *in* Kwongan: plant life of the sandplain ed by J. S. Pate and J. S. Beard. University of Western Australia Press: Nedlands.

Lavery, P. and James, S. H., 1986. Complex hybridity in *Isotoma petraea*. VI. Distorted segregation, gametic lethal systems and population divergence. *Heredity* **58**: 401–08.

Main A. R., 1982. Rare species: precious or dross? Pp. 163–74 *in* Species at risk: research in Australia ed by R. H. Groves and W. D. L. Ride. Australian Academy of Science: Canberra.

Martin, H. A., 1990. Tertiary climate and phytogeography in southeastern Australia. *Paleobotany, Palynology* **65**: 47–55.

Maslin, B. R., 1975. Studies in the genus *Acacia* (Mimosaceae) — 4. A revision of the series Pulchellae. *Nuytsia* **1**: 388–494.

Moran, G. F. and Hopper, S. D., 1983. Genetic diversity and insular population structure of the rare granite rock species *Eucalyptus caesia* Benth. *Aust. J. Bot.* **31**: 161–72.

Moran, G. F. and Hopper, S. D., 1987. Conservation of the genetic resources of rare and widespread eucalypts in remnant vegetation. Pp. 151–62 *in* Nature conservation: the role of remnants of native vegetation ed by D. A. Saunders, G. W. Arnold, A. A. Burbidge and A. J. M. Hopkins. Surrey Beatty & Sons: Sydney.

Morrison, D. A., 1987. The phytogeography, ecology and conservation status of *Lechenaultia* R.Br. (Goodeniaceae). *Kingia* **1**: 85–133.

Ornduff, R., 1986. Comparative fecundity and population composition of heterostylous and non-heterostylous species of *Villarsia* (Menyanthaceae) in Western Australia. *Am. J. Bot.* **73**: 282–86.

Pate, J. S. and Hopper, S. D., 1993. Rare and common plants in ecosystems, with special reference to the South West Australian Flora. In press *in* Ecosystem function of biodiversity ed by E.-D. Schulze and H. A. Mooney. Springer Verlag: Heidelberg.

Sampson, J. F., 1988. The population genetic structure of *Eucalyptus rhodantha* Blakely and Steedman and its allies *Eucalyptus crucis* Maiden and *Eucalyptus lane-poolei* Maiden. Ph.D. Thesis, University of Western Australia: Nedlands.

Sampson, J. F., Hopper, S. D. and James, S. H., 1988. Genetic diversity and the conservation of *Eucalyptus crucis* Maiden. *Aust. J. Bot.* **36:** 447–60.

Sampson, J., Hopper, S. D. and James, S. H., 1989. The mating system and population genetic structure in a bird-pollinated mallee, *Eucalyptus rhodantha. Heredity* **63:** 383–93.

Walker, B. A. and Pate, J. S., 1986. Morphological variation between seedling progenies of *Viminaria juncea* (Schrad. and Wendl.) Hoffmans (Fabaceae) and its physiological significance. *Aust. J. Plant Physiol.* **13:** 305–19.

CHAPTER 3

Patterns of plant functional responses and landscape heterogeneity

R. H. Groves[1] and R. J. Hobbs[2]

The sclerophyll vegetation of the mediterranean regions of southern Australia exhibits many functional responses to low nutrient availability and seasonal water shortages. Plant species can be grouped functionally in relation to their rooting habits and methods of water and energy capture. Vegetation patterning across the landscape results from the interaction of numerous factors, including macro and micro-climate, soil and landform types, disturbance histories and individualistic species distributions. Functional diversity and landscape heterogeneity have both been reduced by human activities.

Introduction

Hopper (Chapter 2) discussed the patterns of diversity of the flora in southwestern Australia at the species and infra-specific levels. In this chapter, we consider other aspects of diversity; namely plant morphological/functional diversity and heterogeneity at the landscape level. We examine the diversity of plant form in sclerophyllous vegetation types in southern Australia as it relates to functional responses to resource limitation and discuss the factors leading to vegetation patterning in mediterranean landscapes.

Functional constraints on vegetation

Taken at its most basic level, the primary function of any vegetation is to photosynthesize. While there is considerable variation in sclerophylly and leaf form shown by a vegetation type such as heath in southern Australia, there is little evidence for different

---

[1]CSIRO Division of Plant Industry, GPO Box 1600, Canberra, Australian Capital Territory, Australia 2601.
[2]CSIRO Division of Wildlife and Ecology, LMB 4, PO Midland, Western Australia, Australia 6056.

Pages 47–60 *in* BIODIVERSITY OF MEDITERRANEAN ECOSYSTEMS IN AUSTRALIA ed by Richard J. Hobbs. Surrey Beatty & Sons, 1992.

photosynthetic groups within the plant community — the majority of plants follow the C_3 pathway of photosynthesis and there are few instances of C_4 or CAM plants. Light rarely limits photosynthesis in any region of mediterranean climate; adaptations to high levels of irradiance may be more significant for ecosystem function. In what follows we shall not consider further the fixation and cycling of carbon in mediterranean-type ecosystems (but see Main, Chapter 5).

To photosynthesize all plants need light, oxygen, water and nutrients. Whereas light may be in excess, shortages of water and nutrients more often limit the growth of sclerophyll vegetation in southern Australia. Mediterranean-type regions are, by definition, limited by regular periods of water shortage through each summer, and so the diversity of response to water shortage will be explored first in this chapter. Another characteristic of such vegetation is its invariable occurrence on soils of very low nutrient status (Specht and Rayson 1957a; Groves 1981). As with water, there is also a diversity of adaptations shown by sclerophyllous shrublands to low nutrient levels (Lamont 1982, Chapter 6). The diversity of plant life present on the very nutrient poor soils present in the mediterranean regions of Australia is remarkable (see Hopper, Chapter 2), and it is possible that this diversity is in part due to the harshness of the environment which does not allow any particular functional response to dominate at the expense of others.

Functional responses to water limitation

A fundamental problem for the growth and survival of the perennial woody plants that dominate sclerophyll vegetation in areas of mediterranean climate is that rain falls mostly in winter (see Hobbs, Chapter 1). Summers are usually hot and dry and surface soil moisture deficits strongly limit plant growth. The plant community responds to this strongly periodic wet-dry cycle by stratification of its shoot and root components, and especially the latter in southern Australian ecosystems. Because most of the dominant species in sclerophyll vegetation tend to grow mainly in summer — especially the proteaceous and eucalypt floristic groups — at a time when temperatures are high but rainfall is low, most emphasis in this section will be placed on adaptations to ensure the availability of soil moisture to the shoot over the summer period. The evolutionary development of such floristic elements in the Tertiary during times of warm-wet (subtropical) summers, long before they had to survive a hot-dry mediterranean-type summer, may be a significant consideration for the discussion which follows (see, e.g., Specht and Rayson 1957a; Raven 1973; Johnson and Briggs 1981).

Grass-trees (*Xanthorrhoea* spp.) are common constituents of heath or the understorey to eucalypt forests and woodlands. Their shoot form especially suits them to the role of water channelers (Fig. 1). Soil moisture is concentrated in the region of the caudex for subsequent uptake and use. Shoots of tall shrubs and trees may also "collect" water and channel it down their stems as "stem-flow" (Specht 1957; Nulsen *et al.* 1986). Soil water storage thus varies according to the variety of shoot form shown by the vegetation and probably reaches its greatest variation in heath vegetation. The above-ground diversity in shoot form, however, is but a small reflection of the greater diversity in form existing below ground in such communities.

Specht and Rayson (1957b) were the first to show the diverse array of root systems of heath vegetation growing on deep sand in southeastern South Australia (Fig. 1). Some species, often the dominants in the community (such as *Banksia ornata*), have deep tap-roots that may penetrate as far as 2 m to the clayey B horizon with a moisture regime very different from that of the sandy top soil. However, Specht and Rayson (1957b) also described two variants of the deep tap-rooted species group — the tap root of one further group decayed with age (e.g., *Casuarina* spp.), whereas the laterals of the other group produced frequent suckering shoots (as in *Banksia marginata* in their community). The same authors described a group of shallow (<0.5 m) tap-rooted species (e.g., *Leucopogon*). Furthermore, most deep tap-rooted species produced extensive lateral root systems as well which develop in surface soil; these fibrous roots arise from different underground organs such as caudices (in the case of *Xanthorrhoea*), stem bases, rhizomes or tubers. All three latter types of fibrous root systems were confined to the top 0.3 m of organically richer soil existing at their South Australian site. Specht and Rayson (1957b) showed that, for the heath they studied, about 25 per cent of the species possessed deep tap roots, about 40 per cent had a shallow tap-rooted system, and the remainder possessed fibrous roots of some sort.

A multiplicity of variants of underground organs was described by Pate and Dixon (1982) for the Western Australian sandplain flora. Their descriptions considerably extend the array of shallow-rooted fibrous root morphs of Specht and Rayson.

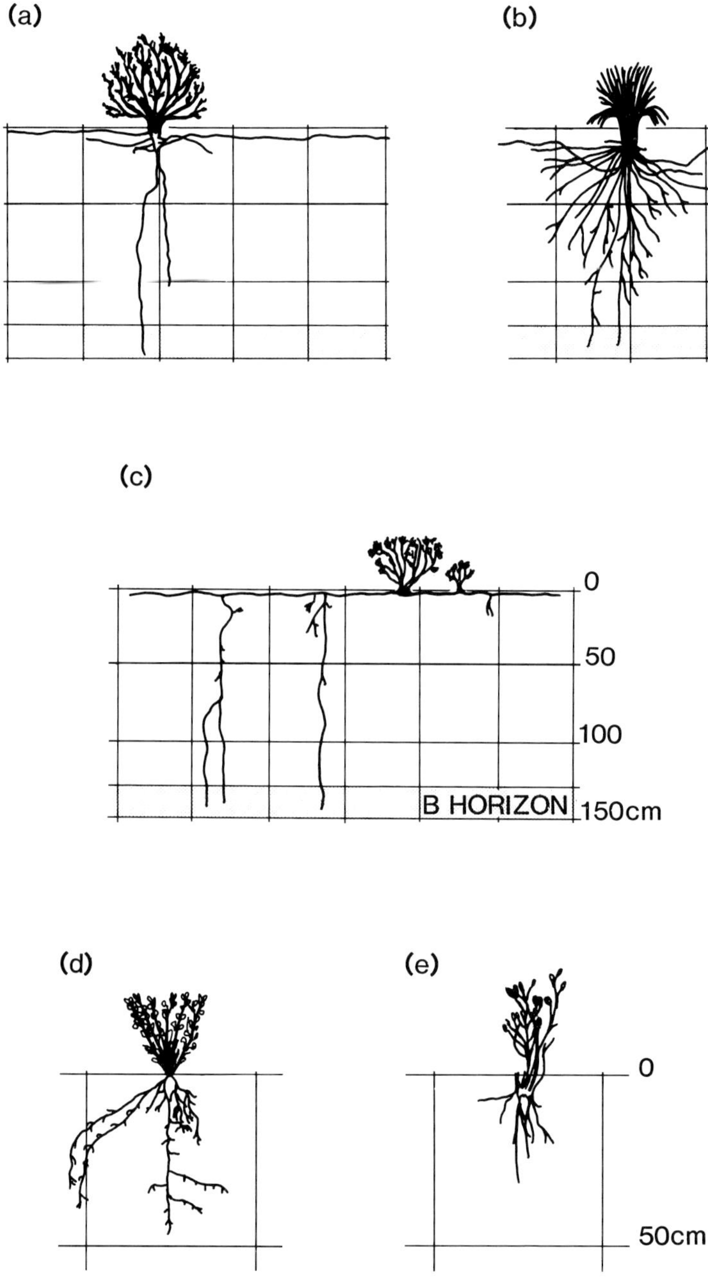
(a)
(b)
(c)
0
50
100
B HORIZON
150cm
(d)
(e)
0
50cm

In summary, a great diversity of root form exists in sclerophyll shrublands in southern Australia, a situation which results in exploitation of soil moisture down the full profile — deep roots in summer tap the soil water stored in the heavier-textured B horizon and lateral fibrous roots tap the sandier A horizon in autumn and spring. A concentration of fine roots at the soil surface and even penetrating the litter layer may also trap scarce moisture coming from the occasional but irregular summer rainstorms that are a feature of most of southern Australia. Sometimes the concentration of surface roots may be most developed in species at the drier limits of their distribution; e.g., as in the distribution of mallee eucalypts being confined increasingly to sand ridges in central Australia as rainfall decreases inland.

Functional responses to nutrient limitation

The diversity in morphological adaptations shown by sclerophyll shrubs and trees is as relevant, if not moreso, to the theme of nutrient exploration and uptake as it is to a consideration of water relations. Deep tap-rooted species may take up not only water but also scarce nutrients from the B horizon and cycle them through the soil profile. The array of forms is even greater with respect to nutrients because of their role in fixing nitrogen, both symbiotically and non-symbiotically, and increasing surface areas for the absorption of nutrients from the less impoverished litter and upper soil layers. This enhanced absorption layer is especially important in the functioning of proteoid roots (Lamont 1982) and mycorrhizae.

One functional group of species that is significantly developed in southern Australian sclerophyll communities is the symbiotic nitrogen fixers, especially the acacias and other legumes. This latter group is not well represented in some comparable vegetation types outside Australia, such as the South African fynbos.

Another functional group within sclerophyll shrublands includes those species which have high levels of nutrients in their seeds, as in obligately-seeding shrubs such as many of the banksias (see Lamont, Chapter 6). This group is important to nutrient cycling in the community, in that seedlings can establish, grow

Fig. 1 (opposite). Main types of root systems in heath vegetation in south-eastern South Australia: (a) *Banksia ornata* (deep tap-root, normal form); (b) *Xanthorrhoea australis* (fibrous roots arising from a stock); (c) *Banksia marginata* (deep tap-root with suckers developing from lateral roots); (d) *Hibbertia sericea* (shallow tap-root); (e) *Hypolaena fastigiata* (fibrous roots arising from stem bases). Redrawn from Specht and Rayson (1957b).

rapidly and take up scarce nutrients soon after fire when, at least in the surface soil, nutrient levels are maximal. Such nutrients can then be recycled as the vegetation matures. Another aspect of the same adaptation is the importance such seeds have in the nutrition of small mammals, and especially soon after a fire, when other nutrient-rich food sources may be in short supply.

Landscape patterns

Ecological heterogenity in the landscape is the result of the interaction of many influences (Kolasa and Rollo 1991; Milne 1991). At a broad scale, patterns of vegetation composition and structure can be related to the rainfall gradients discussed by Hobbs (Chapter 1) and Hopper (Chapter 2), and to changes in soil and landform type (Beard 1983). Generally, forest vegetation is restricted to the higher rainfall areas (> 750 mm), but at lower rainfalls the distribution of woodland, mallee, shrubland and heath vegetation types is determined mainly by soil type (Beard 1983). Thus within any given region, soil type and landform generally determine the broad vegetation patterning (e.g., McArthur 1992). The main vegetation types are also largely characterized by their dominant growth forms (i.e., trees, mallees, shrubs).

Within that broad patterning, however, numerous smaller scales of pattern may be present. These may be determined by finer-scale variations in soil characteristics (e.g., Froend 1987), but may also be the result of other factors (Fig. 2). In particular, the history of disturbances determines the distribution of patches of different age and successional stage across the landscape. The components of the disturbance regime will determine the scale and pattern of variation observed (Pickett and White 1985), and each disturbance type may produce different vegetation responses. Individual disturbance types can also produce different responses depending on factors such as environmental variations within the disturbed areas, weather characteristics following the disturbance, and interactions with other disturbances (Grubb and Hopkins 1986; Armesto *et al.* 1991; Hobbs and Huenneke 1992). Disturbances common in mediterranean ecosystems in Australia include fire,

Fig. 2 (opposite). Diagrammatic illustration of how landscape mosaics develop in response to various factors in mediterranean areas of Australia. The basic pattern in any locality will be determined by the landform and soil types, but superimposed on this will be the pattern of disturbances and individual species distributions. The result is a complex mosaic of vegetation patches of different species compositions, structures and developmental phases.

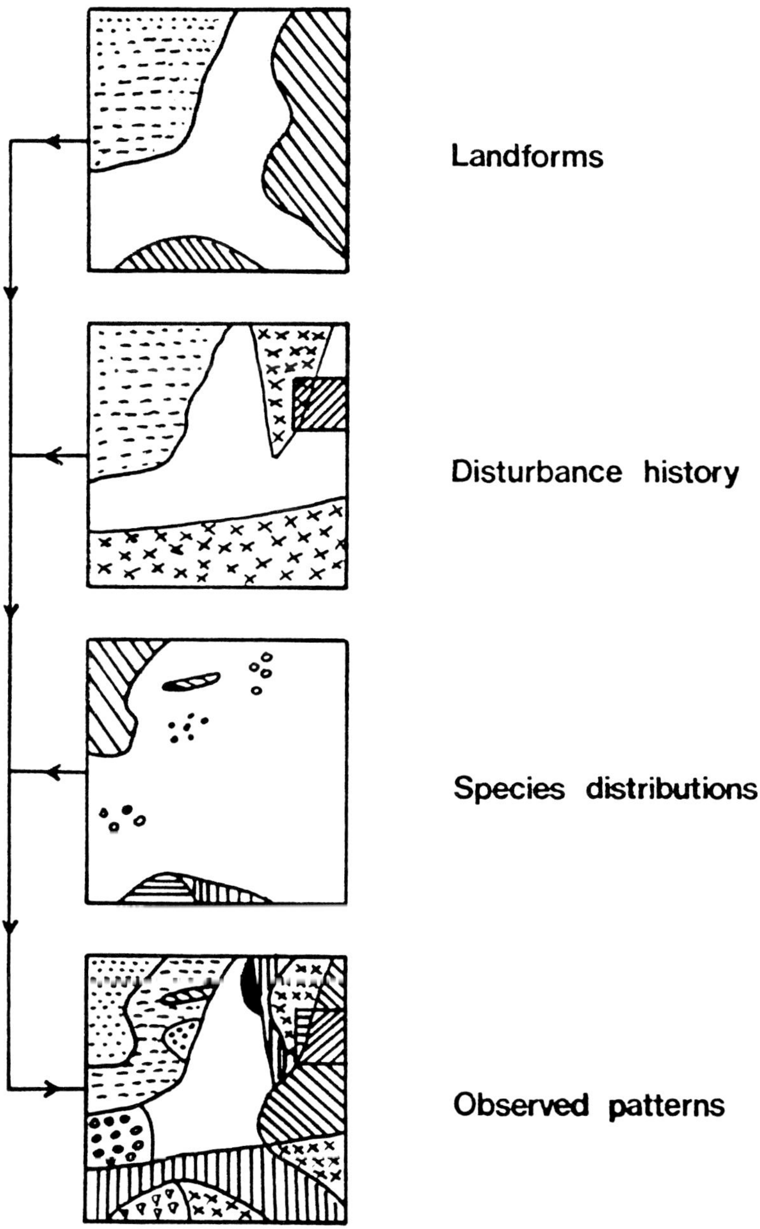
Landforms
Disturbance history
Species distributions
Observed patterns

drought, infrequent frosts or periods of high summer temperatures, severe summer storms, localized soil disturbance by animals, treefalls and insect outbreaks (Hnaitiuk and Hopkins 1980; Grubb and Hopkins 1986; van Schagen *et al.* 1992; Main, Chapter 5). In addition to the natural disturbance regime, human disturbance is now an important component of most mediterranean ecosytems. Human activities such as vegetation removal, timber or soil extraction and introduction of non-native grazers produce a further overlay of variation on the landscape, as well as altering the natural regime (Hobbs 1987; Scougall *et al.* 1993; Saunders and Hobbs, Chapter 4).

High levels of species turnover across the landscape, as described by Hopper (Chapter 2) and patchy distribution of populations of individual species can also add to the complexity of vegetation patterning. This patchiness can be as a result of chance dispersal events, response to localized disturbances or response to microenvironmental variation (Weiss and Murphy 1993). As an example of species response to microenvironmental variation, Rayson (1957) mapped the distribution of plant species across a crescentic dune at Dark Island heath in southeastern South Australia. The distribution of many species varied along a west-east transect across the dune surface. Depth to the surface of the B horizon varied greatly with position across the dune. So too did rainfall amount and soil particle size. With predominantly westerly winds, rainfall was lowest on the west-facing slope and highest on the dune-slip face and the sandplain (an effect opposite to that of rain-shadows created by mountain ranges). The interaction between sand particle size and higher rainfall was shown to influence significantly the distribution of clumps of *Eucalyptus baxteri* (Fig. 3). This species, a mallee eucalypt and the only "tree" at this site, was present only on the east-facing, wetter slope. The role of microtopography in determining the distribution of soil particles and rainfall incidence is clearly significant for botanical composition in this instance (see also Lamont, Chapter 6).

Changes following European colonization

In southern Australia many of the areas where such sclerophyll shrublands existed previously are now cleared of indigenous vegetation (see Saunders and Hobbs, Chapter 4) and sown to agricultural species, such as grasses, legumes and cereals, all of Mediterranean origin. The growth of most of these introduced species is confined to the autumn-spring period because of their predominantly shallow-rooting habit. Such introduced species fail

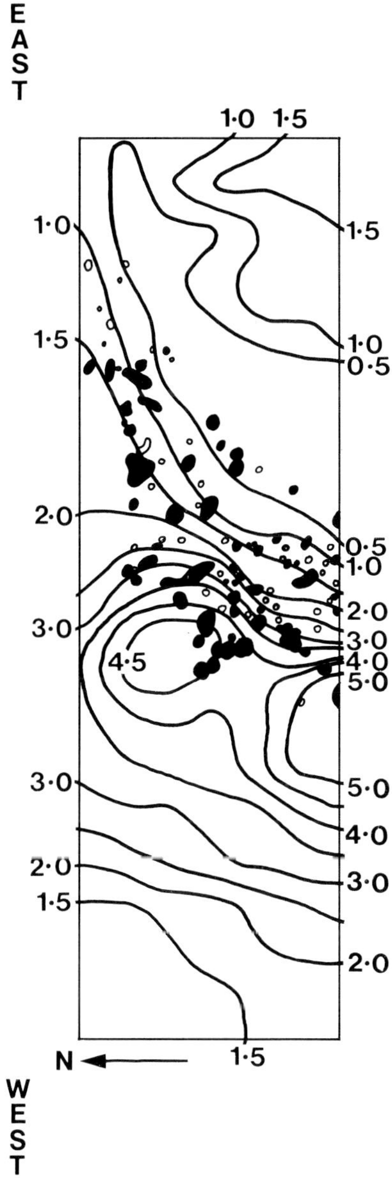

Fig. 3. Map of position of a crescentic W–E sand dune showing the distribution of clumps of *Eucalyptus baxteri* in relation to altitude above the sandplain. 1 unit = 3.0 m. Redrawn from Rayson 1957.

to tap the soil moisture stored in the B horizon, with the exception of a few deeper-rooted Mediterranean species such as lucerne (*Medicago sativa*) and phalaris (*Phalaris aquatica*). It is significant that many weeds also introduced from the Mediterranean Basin are mainly shallow-rooted annuals (Groves 1986; Fox 1990). Only a few deep-rooted perennials such as skeleton weed (*Chondrilla juncea*) are effective summer-green species because they too tap soil moisture stored in the B horizon. In this respect, weeds such as skeleton weed and agronomic species such as lucerne "mimic" the adaptations to low moisture levels in surface soil in summer shown by indigenous vegetation.

The undisturbed ground surface beneath the sclerophyll shrubs is typically covered by small perennial forbs, lichens, a few grasses, many graminoids (depending on soil water status) and a layer of litter that decomposes slowly. The latter component of the ground layer is especially prominent under mallee eucalypts. With the introduction of domesticated grazing animals and a change in the burning regime that accompanied such grazing, the disturbance regime changed radically. The overall result of these changes has been to shift the floristic composition of the understorey from a suite of perennial species to one composed mainly of annual herbs, most of which are introduced species. There has been a loss of surface soil particles and nutrients and a consequent decrease in ecosystem stability. The loss of lichen cover from the ground layer may be of especial significance to ecosystem function and is worthy of research attention.

A major ecological consequence of replacement of deep-rooted trees by shallow-rooted herbs has been an enhanced movement of salt through the soil profile and resultant salinization of surface water and agricultural land (Peck 1978; see also Saunders and Hobbs, Chapter 4). In addition, increased salinity may affect the surface and groundwater supplies on which the city of Perth depends. Only retention of remaining native vegetation (in good condition) coupled with strategic replanting of salt-tolerant tree species can prevent increasing salinization of drinking water supplies (Bell 1988). At a larger scale, the entire region of south-eastern Australia is under threat from salinization. Extensive replacement of indigenous tree cover by introduced pasture and crop plants in the catchments of the Murray and Darling Rivers, together with continued development of irrigation schemes along the Murray River for the last 100 years, have led to local and regional increases in salt levels in both river and ground-water that is used for human and domesticated animal water supply throughout the region (Brett 1990). Below the large catchment system for surface water lies an equally large interconnected groundwater system, the quality of which is increasingly at risk because of the ecological consequences of previous land-use changes, especially the clearance of trees. These effects are displaced in both time and space but are increasingly affecting ecosystem function at the supra-regional scale of southeastern Australia. Clearly, salinization is one manifestation of altered ecosystem function following human activities, and more general changes in function are also taking place, as discussed in other chapters.

Conclusions

Examining the functional responses of plant species to environmental constraints is one way of simplifying the vast diversity with which we are confronted in mediterranean Australia (as discussed by Hopper, Chapter 2). We have not attempted an exhaustive list of plant functional groups, since the development of such groups is still in its infancy in this area (but see Leishman and Westoby 1992). Rather we have illustrated the diverse array of morphological types which are present in mediterranean schlerophyll vegetation and which probably have functional significance in dealing with water and nutrient limitation.

In addition, we have discussed the vegetation patterning in mediterranean landscapes, and have illustrated that complex vegetation mosaics result from the interplay of environmental variation at a broad and fine scale, disturbances of various types, and individualistic species distributions.

Does functional diversity and landscape complexity matter? Certainly, it seems clear that the change from indigenous to introduced vegetation has had, and continues to have, profound effects on ecosystem function in southern Australia. This change has generally been from perennial to annual, from deep-rooted to shallow-rooted and from species-rich to species-poor (i.e., crop and pasture) plant communities. The options available for water and nutrient capture in a heterogeneous and uncertain environment have thus been reduced. Reduction in landscape complexity also has implications for the conservation of species diversity. It is likely that some species require certain combinations of disturbance and environmental conditions for establishment and/or survival. Animal species may also be dependent on vegetation heterogeneity (e.g., Saunders and Ingram 1987; Lambeck and Saunders 1993). Clearly, therefore, these aspects of diversity are important over and above considerations of species diversity *per se*. Conservation in the mediterranean regions of Australia thus needs to consider functional and landscape diversity, as well as the genetic diversity discussed by Hopper (Chapter 2). A species-oriented approach is no longer acceptable, and overall ecosystem and landscape management will be required for the retention of biodiversity.

Acknowledgements

We thank Lyn Atkins for preparing the figures and Denis Saunders for critical review of the manuscript.

References

Armesto, J. J., Pickett, S. T. A. and McDonnell, M. J., 1991. Spatial heterogeneity during succession: a cyclic model of invasion and exclusion. Pp. 256–69 *in* Ecological heterogeneity ed by J. Kolasa and S. T. A. Pickett. Springer-Verlag: New York.

Beard, J. S., 1983. Ecological control of the vegetation of southwestern Australia: moisture versus nutrients. Pp 66–73 *in* Mediterranean-type ecosystems: The role of nutrients ed by F. J. Kruger, D. T. Mitchell and J. U. M. Jarvis. Springer-Verlag: Berlin.

Bell, A., 1988. Trees, water and salt — a fine balance. *Ecos* **58**: 2–8.

Brett, D., 1990. Restoring the Murray-Darling Basin. *Ecos* **64**: 4–9.

Fox, M. D., 1990. Mediterranean weeds: exchanges of invasive plants between five Mediterranean regions of the world. Pp. 179–200 *in* Biological invasions in Europe and the Mediterranean basin ed by F. di Castri, A. J. Hansen and M. Debussche. Kluwer: Dordrecht.

Froend, R. H., 1987. Investigations into species richness patterns in the northern sandplain region of Western Australia. Ph.D. Thesis, University of Western Australia: Nedlands.

Groves, R. H., 1981. Heathland soils and their fertility status. Pp. 143–50 *in* Ecosystems of the World 9B. Heathlands and related shrublands, analytical studies ed by R. L. Specht. Elsevier: Amsterdam.

Groves, R. H., 1986. Invasion of mediterranean ecosystems by weeds. Pp. 129–45 *in* Resilience in mediterranean-type ecosystems ed by B. Dell, A. J. M. Hopkins and B. B. Lamont. Junk: Dordrecht.

Grubb, P. J. and Hopkins, A. J. M., 1986. Resilience at the level of the plant community. Pp. 21–38 *in* Resilience in mediterranean-type ecosystems ed by B. Dell, A. J. M. Hopkins and B. B. Lamont. Junk: Dordrecht.

Hnatiuk, R. J. and Hopkins, A. J. M., 1980. Western Australian species-rich kwongan (sclerophyllous shrublands) affected by drought. *Aust. J. Bot.* **28**: 573–85.

Hobbs, R. J., 1987. Disturbance regimes in remnants of natural vegetation. Pp. 233–40 *in* Nature conservation: the role of remnants of native vegetation ed by D. A. Saunders, G. W. Arnold, A. A. Burbidge and A. J. M. Hopkins. Surrey Beatty & Sons: Chipping Norton, NSW.

Hobbs, R. J. and Huenneke, L. F., 1992. Disturbance, diversity and invasion: implications for conservation. *Conserv. Biol.* **6**: in press.

Johnson, L. A. S. and Briggs, B. G., 1981. Three old southern families — Myrtaceae, Proteaceae and Restionaceae. Pp. 427–69 *in* Ecological biogeography of Australia Vol. 1. ed by A. Keast. Junk: The Hague.

Kolasa, J. and Rollo, C. D., 1991. Introduction: the heterogeneity of heterogeneity: a glossary. Pp. 1–23 *in* Ecological heterogeneity ed by J. Kolasa and S. T. A. Pickett. Springer-Verlag: New York.

Lambeck, R. J. and Saunders, D. A., 1993. The role of patchiness in reconstructed wheatbelt landscapes. In press *in* Nature conservation 3: reconstruction of fragmented ecosystems, global and regional perspectives ed by D. A. Saunders, R. J. Hobbs and P. R. Ehrlich. Surrey Beatty & Sons: Chipping Norton, NSW.

Lamont, B. B., 1982. Mechanisms for enhancing nutrient uptake in plants, with particular reference to mediterranean South Africa and Western Australia. *Bot. Rev.* **48:** 597–689.

Leishman, M. R. and Westoby, M., 1992. Classifying plants into groups on the basis of association of individual traits — evidence from Australian semi-arid woodlands. *J. Ecol.* **80:** in press.

McArthur, W. M., 1992. History of landscape development. In press *in* Reintegrating fragmented landscapes: towards sustainable production and nature conservation ed by R. J. Hobbs and D. A. Saunders. Springer-Verlag: New York.

Milne, B. T., 1991. Heterogeneity as a multiscale characteristic of landscapes. Pp. 69–84 *in* Ecological heterogeneity ed by J. Kolasa and S. T. A. Pickett. Springer-Verlag: New York.

Nulsen, R. A., Bligh, K. J., Baxter, I. N., Solin, E. J. and Imrie, D. H., 1986. The fate of rainfall in a mallee and heath vegetated catchment in southern Western Australia. *Aust. J. Ecol.* **11:** 361–71.

Pate, J. S. and Dixon, K. W., 1982. Tuberous, cormous and bulbous plants. University of Western Australia Press: Perth.

Peck, A. J., 1978. Salinization of non-irrigated soils and associated streams: A review. *Aust. J. Soil. Res.* **16:** 157–68.

Pickett, S. T. A. and White, P. S., 1985. The ecology of natural disturbance and patch dynamics. Academic Press: New York.

Raven, P. H., 1973, The evolution of Mediterranean floras. Pp. 213–24 *in* Mediterranean type ecosystems ed by F. di Castri and H. A. Mooney. Springer-Verlag: Berlin.

Rayson, P., 1957 Dark Island heath (Ninety-Mile Plain, South Australia). II. The effects of microtopography on climate, soils and vegetation. *Aust. J. Bot.* **5:** 86–102.

Saunders, D. A. and Ingram, J. A., 1987. Factors affecting survival of breeding populations of Carnaby's cockatoo *Calyptorhynchus funereus latirostris* in remnants of native vegetation. Pp. 249–58 *in* Nature conservation: the role of remnants of native vegetation ed by D. A. Saunders, G. W. Arnold, A. A. Burbidge and A. J. M. Hopkins. Surrey Beatty & Sons: Chipping Norton, NSW.

Scougall, S. A., Majer, J. D. and Hobbs, R. J., 1993. Edge effects in grazed and ungrazed Western Australian wheatbelt remnants in relation to ecosystem reconstruction. In press *in* Nature conservation 3: reconstruction of fragmented ecosystems, global and regional perspectives ed by D. A. Saunders, R. J. Hobbs and P. R. Ehrlich. Surrey Beatty & Sons: Chipping Norton, NSW.

Specht, R. L., 1957. Dark Island heath (Ninety-Mile Plain, South Australia). IV. Soil moisture patterns produced by rainfall interception and stemflow. *Aust. J. Bot.* **5**: 137–50.

Specht, R. L. and Rayson, P., 1957a. Dark Island heath (Ninety Mile Plain, South Australia). I. Definition of the ecosystem. *Aust. J. Bot.* **5**: 52–85.

Specht, R. L. and Rayson, P., 1957b. Dark Island heath (Ninety Mile Plain, South Australia). III. The root systems. *Aust. J. Bot.* **5**: 103–114.

van Schagen, J. J., Hobbs, R. J. and Majer, J. D., 1992. Defoliation of trees in roadside corridors and remnant vegetation in the Western Australian wheatbelt. *J. Roy. Soc. West. Aust.* **75**: in press.

Weiss, S. and Murphy, D., 1993. Climatic considerations in reserve design and ecological restoration. In press *in* Nature conservation 3: reconstruction of fragmented ecosystems, global and regional perspectives ed by D. A. Saunders, R. J. Hobbs and P. R. Ehrlich. Surrey Beatty & Sons: Chipping Norton, NSW.

CHAPTER 4

Impact on biodiversity of changes in land-use and climate

D. A. Saunders[1] and R. J. Hobbs[1]

The mediterranean climatic zones of Australia occur in the south-west of Western Australia and in southeastern South Australia and western Victoria. These are major centres of extensive cereal cropping and sheep farming. Widespread clearing of native vegetation for agriculture, introduction of feral animals and inappropriate agricultural methods have led to land degradation and loss of native species of flora and fauna. Over-exploitation of old growth, changes in fire regime and introduced pathogens have also markedly altered the forest ecosystems of southwestern Australia.

Historical data indicate that the climate of the region, particularly rainfall, is very variable. There appears to have been an overall reduction in winter rainfall in southwestern Australia during this century, but the data on this are not conclusive. Detecting future changes due to changed atmospheric composition will be very difficult due to the inherent climatic variability. We argue that past and future changes in land-use are likely to be of much greater importance than climatic changes in their effect on biodiversity. Land-use changes thus require greater emphasis than currently accorded in discussions of global change.

Introduction

We live in a time of rapid and extensive change. At the global level, it seems certain that the population of *Homo sapiens* will continue to rise and resource demand will continue to increase around the world. Ehrlich and Ehrlich (1990, 1991) note that the human population was about 5–10 million at the time of the agricultural revolution and had increased to 5.4 billion by 1991. They also note that the average impact per person (using energy consumption as a measure of impact) has increased about 8 times.

---

[1]CSIRO Division of Wildlife and Ecology, LMB 4, PO Midland, Western Australia, Australia 6056.

Pages 61–75 *in* BIODIVERSITY OF MEDITERRANEAN ECOSYSTEMS IN AUSTRALIA ed by Richard J. Hobbs. Surrey Beatty & Sons, 1992.

These interconnected factors are already having a major impact on ecosystems and this impact shows no signs of abatement. There is no ecosystem in the world that has not been affected by the "scale of the human enterprise" (Ehrlich 1993a). We can state with certainty that this accelerated growth in human population will continue, as will demand for resources to allow a better standard of living, or at least to retain the *staus quo*. However, we are uncertain as to what the result of the increase will be. It is highly likely that our present rate of growth will result in widespread loss of biotic diversity, life support systems and, possibly, the eventual demise of *H. sapiens*. This is likely even without the potential human-induced changes in climate which are predicted (e.g., Henderson-Sellers and Blong 1989; Schneider 1989).

In Australia, over the last 200 years there have been enormous changes wrought on the continent by European settlers and their descendants exploiting the land and its biota (Saunders *et al.* 1990; Cocks 1992; Hobbs and Saunders 1992). The extent of these changes varies between different climatic zones across the continent, but, they are undoubtedly greatest in the intensively farmed areas of Australia, particularly those within the mediterranean climatic zones. Here we examine the impacts of these changes on the biodiversity of the mediterranean regions, with particular emphasis on southwestern Australia. We also consider the current predictions of climatic changes in the region and assess their likely impact on biodiversity.

Impact of land-use on agricultural productivity

The Western and southeastern Australian mediterranean environments (see Hobbs, Chapter 1) comprise nearly half of the cereal cropping areas of Australia and are Western and South Australia's agricultural heartlands. We will concentrate on southwestern Australia to discuss the impact of land-use and climate change on biodiversity in Australian mediterranean environments. However, the situation in southeastern Australia is almost certainly the same, although some of the details will undoubtedly differ (Boardman 1986; van der Sommen 1986; Taylor 1987).

The wheatbelt of Western Australia has 15.6 million ha of cleared land which, in 1989/90, yielded agricultural products worth 3.5 billion dollars (Nulsen 1992). Over half of this was from crops, the remainder coming from livestock products, particularly wool.

Unfortunately this production has not been achieved without a cost to the land. For example, nearly 3 per cent of cleared land is affected by soil salinity at present, and the predictions are that over 15 per cent may be affected possibly within the next 30 years (Nulsen 1992) (Fig. 1a). At present, 13 per cent of cleared land is affected by wind erosion, over 54 per cent by subsoil compaction, and 11 per cent by the decline in soil structure which can reduce productivity by at least 50 per cent. Evidently soil structure decline has the potential to influence an area twice that currently affected (Nulsen 1992).

There is little doubt that the agricultural enterprises being practiced in the agricultural areas of Western Australia are not sustainable. As most of the area of land in southeastern Australia which is under the influence of mediterranean climates also suffers from moderate to severe land degradation (Young *et al.* n.d.; Cocks 1992) the situation is almost certainly the same there. At the moment, our agricultural soils are degrading and this may lead to a long-term, and probably permanent, decrease in productivity. It does not take much to realize that with the increasing global human population, the land will be asked to yield more and may thus be further degraded; this is without any change in climate or current landuse.

Impact of land-use on biodiversity

The flora of the mediterranean climatic zone of Western Australia is very diverse (Pate and Beard 1984), but the vegetation has suffered a dramatic reduction in area since clearing for agriculture began (Fig. 1b,c), with concomitant effects on its associated fauna. Forest regions, while not suffering from the same decline in area, have been subjected to widespread logging, altered fire regimes and introduced pathogens, all of which have altered forest structure and composition (see Hobbs, Chapter 1). Hobbs *et al.* (1992) have reviewed the changes in the native biota of the agricultural area which have resulted from land clearing and associated disturbances and a summary of these changes follows.

After Europeans settled Western Australia, they moved inland and grazed domestic livestock on native vegetation. This had an impact on native species, particularly around watering points where pastoral activities tended to be concentrated (Hobbs *et al.* 1992). Then the agricultural phase began. Over 93 per cent of the original vegetation was removed during this phase (Fig. 1b,c; see also Hobbs, Chapter 1; Majer, Chapter 9), with over 55 per cent being removed after 1945. Because vegetation types are closely allied to particular soil types, some plant communities were much more extensively modified than others. Woodlands on heavier soils, in particular, were cleared and utilized preferentialy. The result is the known extinction of 24 species of plant (Leigh *et al.* 1984) and the possible unknown loss of many others. The agricultural area of Western Australia now has the highest numbers of rare and endangered plant species in Australia (Hartley and Leigh 1979; Leigh *et al.* 1984; Hopper *et al.* 1990). Because of selective clearing of favoured soil types, woodland species are poorly represented in the remaining native vegetation (Beard and Sprenger 1984) and, unless land management changes, are endangered throughout the agricultural area. Clearing thus not only changed the abundance of native vegetation, but also changed the patterns and availability of the different vegetation types present.

Grazing and disturbance by stock and rabbits within remaining vegetation patches further alters the vegetation and results in changed soil composition and structure. Introduced plant species have spread into remnant areas where they can affect vegetation composition and regeneration (Hobbs and Atkins 1991; Scougall *et al.* 1993).

Removal of the native vegetation and the introduction of alien species (particularly rabbits and foxes) has resulted in the depletion of the native fauna. Kitchener *et al.* (1980) noted that only 12 of the 43 species of native mammal recorded from the

Fig. 1. (a) Land affected by soil salinity. *(b and c)* Aerial obliques of central wheatbelt to show clearing pattern.

wheatbelt at European settlement were still common. In fact, nine of the original 43 species are now extinct on the Australian mainland, and mammals are still being lost from the wheatbelt (Hobbs *et al.* 1992).

The avifauna is currently demonstrating the same pattern of species decline as the mammals (Saunders 1989; Saunders and Curry 1990; Saunders 1992). For example, of the 192 species of birds recorded from the agricultural areas since the turn of the century, 50 per cent have declined in distribution and/or abundance and at least three species have disappeared from the area altogether (Saunders 1992). If current land-use continues, this number will certainly rise.

There is little information on the changes in status of other vertebrates as a result of development for agriculture. There is even less information on the effects on invertebrates. The few studies available indicate that development for agriculture and subsequent agricultural practices have reduced the diversity of the soil macrofauna and significantly altered its effects on soil composition, nutrient cycling and water movement (Abbott *et al.* 1979; Abbott and Parker 1980; Lobry de Bruyn 1990; Hobbs *et al.* 1992). There are also few data available on changes in soil microfauna and flora and the potential impacts of these changes on pedological and other processes.

There is no doubt that the changes to our native biota and ecosystem processes as a consequence of our development of the agricultural area are resulting in a rapid loss of biodiversity. In some groups, this loss is measureable and shows no signs of slowing, while information on many groups is still sparse or lacking. One of the principles of sustainability espoused in the document "Caring for the earth. A strategy for sustainable living" (WCU/UNEP/WWF 1991) is the maintenance of biological wealth. This has two goals, namely:

1. To maintain life support services, or the ecological processes that sustain the productivity, adaptability and capacity for renewal of lands, waters and all life on earth.
2. To conserve the variety of life in all its forms.

Clearly, neither of these goals is being realized in much of the agricultural area of southwestern Australia. The same is broadly true in the forest region (see Hobbs, Chapter 1). In emotive terms, we are mining our soils and failing to establish a sustainable conservation system. Past land-use has had a dramatic influence on the biodiversity of the region, and current practices have, until recently, done little to reduce this influence.

Changes in climate

At present there is considerable interest at all levels in "the greenhouse effect" and resulting climate change (Lowe 1989; Pearman 1988). This extends to the popular media, and articles are common in newspapers and on radio and television. There is a great deal of concern over potential rapid changes in climate over the next 50 years. However, we have no clear idea of how, or if, climates have been changed over the last 100 years since clearing began in the wheatbelt of Western Australia. The inherent variability of rainfall in the area is illustrated in Figure 2. With three-fold differences in rainfall amounts between wettest and driest years, it is very difficult to detect any trends or nonlinearities over the period for which records are available. This variability also makes it very difficult to detect any changes due to human influences on atmospheric composition. Pittock (1988) presented data for southwestern Australia which express the trends in winter rainfall as per cent per decade between 1913 and 1986 (Fig. 3). The analysis demonstrated that the northern wheatbelt had a 4.8 per cent drop in winter rainfall, the central wheatbelt a 4.0 per cent drop, and the southern wheatbelt a 3.6 per cent drop. In contrast, the goldfields east of the agricultural area demonstrated a 5.7 per cent increase over the same period. Pittock (1988) noted that "The apparently upward trend in the district around Kalgoorlie may be due to inhomogeneities in the data, especially as there are only a few stations in the district". However, an alternative interpretation is that changes in the radiation balance due to clearing of perennial native vegetation and its replacement by annual crops results in reduced cloud formation over the wheatbelt, resulting in more moisture being carried inland. Current studies of satellite images, although not complete, suggest that cloud formation in the wheatbelt is greatest over the few remaining large blocks of native vegetation (R. Smith, pers. comm.). Hence land-use and climate changes may be interlinked in ways previously not considered.

There is a great deal of uncertainty regarding the future for agricultural production and conservation of biodiversity in Australia's mediterranean climatic zones under the present climatic regime (Hobbs and Saunders 1992). However, with rising concentrations of CO_2 and other greenhouse gases in the atmosphere, it is

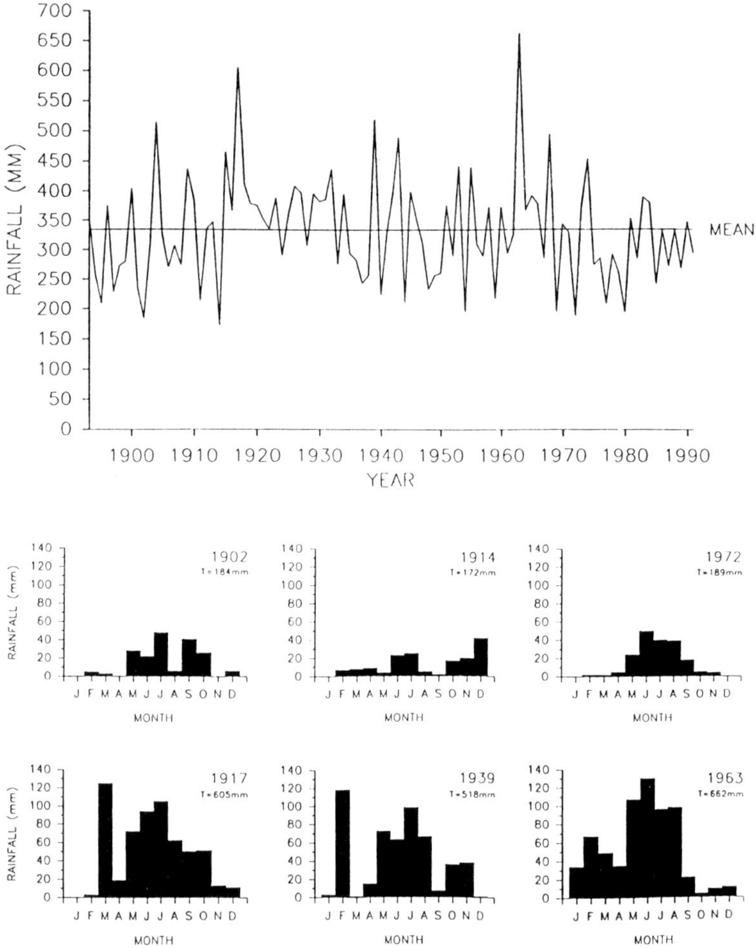

Fig. 2. (a) Annual rainfall (mm) for Kellerberrin, Western Australia over the period 1893-1991, and *(b)* a comparison of monthly totals during the three years of lowest rainfalls (1902, 1914 and 1972) with the three years of highest rainfalls (1971, 1939, 1963). (T = total annual rainfall). Data from Bureau of Meteorology, Perth.

highly likely that these zones will face changes in climate which may be considerably more rapid than have occurred in the past (Pearman 1988). There is a high degree of uncertainty associated with the discussion of climate change relative to the certainty of other changes with which we are faced at the moment. In addition, it seems that a great deal of effort and money is being spent on the debate on climate change, and in planning for climate change at a time when we are incapable of planning to manage the changes we have already imposed on Australian ecosystems.

Climate is the expression of very complex interactions between the energy of the sun, and land masses, bodies of water, the atmosphere and living organisms, particularly plants. Because of this complexity we do not have the data nor the ability to compute general circulation models that are capable of predicting

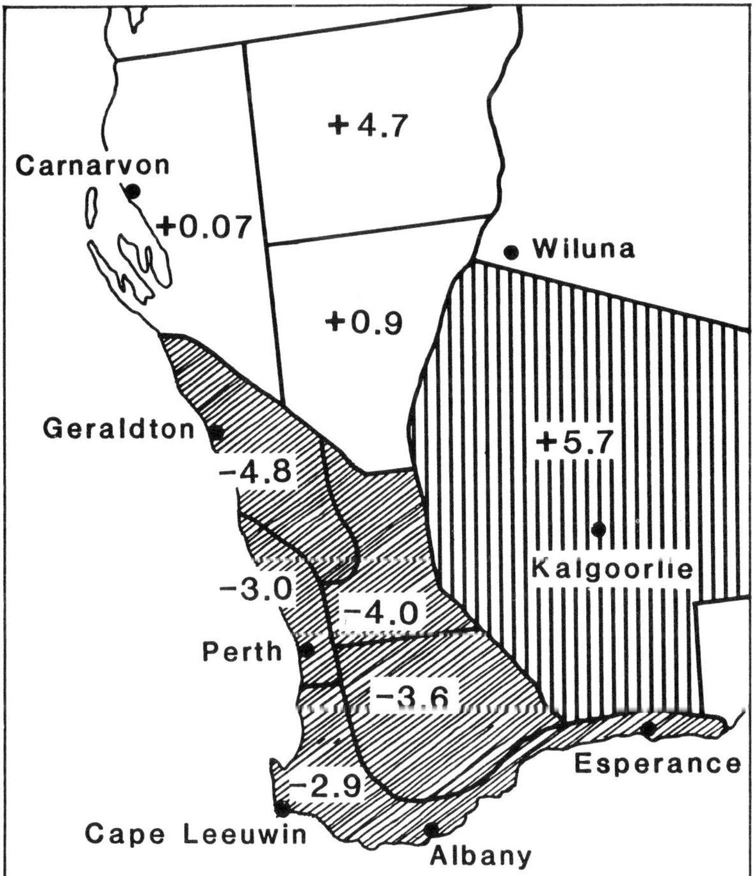

Fig. 3. Trends in winter rainfall in per cent per decade, between 1913–1986 for southwestern Australia (redrawn from Pittock 1988, Fig. 4).

present climate with accuracy, let alone predicting climate change with certainty (Pearman 1988). In fact, predictions concerning the future climate of the southwestern Australian agricultural area illustrate this lack of certainty.

Two scenarios, based on differing initial assumptions, for climate change were suggested by Hille (1989) for southwestern Australia. The first scenario predicted rises in temperatures of the order of 1.1–1.4°C, a decrease in winter rainfall in the south-west and up to a 40 per cent increase in summer rainfall. The decrease in winter rainfall is likely to be most pronounced (up to 20%) in the western parts of the south-west. Under a second scenario, temperatures would rise by 3.2–5.6°C with the likelihood of a slight increase in rainfall in the south-west in winter and up to a 40 per cent increase in summer rainfall. In both scenarios winters would be milder than at present and summers hotter in the first scenario or considerably hotter in the second. Hille's scenarios were modelled to predict the change by 2040 (i.e., in 50 years). When we consider the size of these changes and the rapidity of the changes in relation to past changes, we are considering changes as much as 100 times faster than have occurred before (Hobbs and Hopkins 1991).

There are vastly different inferences to be drawn from Hille's (1989) scenarios. One implies drier winters (and hence cereal crop growing seasons), and the other slightly wetter winters with summers very much hotter than present. As pointed out by Main (1993), Australian mediterranean environments are bounded towards the tropics by hotter, drier semi-arid and arid regions and in the south by oceans. Thus, species confined to these zones are in no position to move south or west, and species dependent on moist areas are in no position to move inland under either of Hille's (1989) scenarios: a variation of being caught between a rock and a hard place.

Recent discussions of the implications of global climate change for natural ecosystems and biodiversity can be found in Clark (1991), Peters (1991) and Lester and Myers (1991). Wyman (1991) also discusses how climate change is likely to interact with habitat fragmentation. Main (1988, 1993) and Arnold (1988) have discussed the potential impact of one scenario of climate change on the biota of southwestern Australia, but both emphasized the uncertainties involved. Hobbs and Hopkins (1991) have reviewed the problem of predicting biotic response to climate change and concluded that simplistic models of geographic shifts of species and communities were of little value. We have a poor level of understanding of the functioning of natural ecosystems in Australian mediterranean regions under present climatic conditions, far less under rapidly changing conditions. Any predictions are therefore little more than informed guesswork at present.

Discussion

The major conclusion we draw is that the most significant impact on biodiversity in the mediterranean regions of Australia has been the rapid changes in land-use over the past century. The continuing impact of clearing for agriculture in terms of habitat loss, fragmentation and land degradation is also likely to outweigh any potential impact of climate change over the next 50 years. Further to this, it may be next to impossible to actually detect any effects of climate change on the biota because they are liable to be confounded by the continuing effects of fragmentation and disturbance.

Land use changes can be expected in the future, and the composition of the atmosphere will continue to change as more greenhouse gases are pumped into the atmosphere. We are uncertain as to the effects, but we may be in for a period of unprecedented change as rapid shifts in climate are superimposed on the effects of landscape fragmentation and degradation. The uncertainty makes it difficult to plan for the conservation of the biota in Australia's mediterranean environments. This difficulty should not be an excuse to do nothing.

There are encouraging signs that rural communities are responding to their degrading landscapes and trying to reverse the degrading processes. The very rapid formation of many Land Conservation District Committees throughout Western Australia is a sign of this response (Goss and Chatfield 1993). However, the problems associated with restoring ecological systems in the mediterranean climatic zone of Western Australia may be exacerbated because during the restoration phase the region may develop a climate which is significantly different from present (Main 1993). For example, if the region is hotter and drier many of the species adapted to cooler more moist conditions will have no suitable environment (Main 1993). There is a need to develop revegetation models which combine nature conservation with soil conservation and which can be put into effect at the landscape scale so that potential future changes can be taken into account (Hobbs and Saunders 1991, 1992; Lefroy *et al.* 1991).

Without such developments we cannot manage our agricultural landscapes to prevent the loss of biodiversity and the degradation of the ecological processes that sustain that biodiversity, far less develop strategies to cope with climate change. Hobbs and Hopkins (1991) suggested that "We may have little hope of preserving all but a vestige of the diversity that still exists if climatic changes are as rapid as some models predict, but we will have even less hope if we do not do something soon." Can we produce a response which is general enough to allow us to start managing our natural resources and agricultural production given high degrees of uncertainty, and over a short timeframe? Given what we have done to our environment now and our limited reponse to those detrimental changes, we can be excused for being pessimistic about our ability to develop a response and act on it. Main (Chapter 10) provides one framework for developing decision strategies to cope with the high levels of uncertainty involved. We suggest that such strategies need to be developed urgently. As Ehrlich (1993b) has pointed out, we do not have the luxury of waiting until we can be more certain; research and action have to go hand in hand.

Acknowledgements

We thank Graham Arnold, Jonathan Majer and Bert Main for constructive comments on the manuscript.

References

Abbott, I. and Parker, C. A., 1980. Agriculture and the abundance of soil animals in the Western Australian wheatbelt. *Soil Biol. Biochem.* **12:** 455–59.

Abbott, I., Parker, C. A. and Sills, I. D., 1979. Changes in abundance of large soil animals and physical properties of soils following cultivation. *Aust. J. Soil Res.* **17:** 343–53.

Arnold, G. W., 1988. Possible effects of climatic change on wildlife in Western Australia. Pp. 375–86 *in* Greenhouse. Planning for climate change ed by G. I. Pearman. CSIRO: Melbourne.

Beard, J. S. and Sprenger, B. S., 1984. Geographical data from the vegetation survey of Western Australia. Vegetation Survey of Western Australia Occasional Paper 2. Vegmap Publications: Perth.

Boardman, R., 1986. The management of South Australia's forests and woodlands. *In* The ecology of the forests and woodlands of South Australia ed by H. R. Wallace. D. J. Woolman Government Printer: Adelaide.

Clark, J. S., 1991. Ecosystem sensitivity to climate change and complex responses. Pp. 65–98 *in* Global climate change and life on earth ed by R. L. Wyman. Routlidge, Chapman and Hall: New York.

Cocks, D., 1992. Use with care. New South Wales University Press: Kensington.

Ehrlich, P. R., 1993a. The scale of the human enterprise. In press *in* Nature conservation 3: reconstruction of fragmented ecosystems, global and regional perspectives ed by D. A. Saunders, R. J. Hobbs and P. R. Ehrlich. Surrey Beatty & Sons: Chipping Norton, NSW.

Ehrlich, P. R., 1993b. Biodiversity and ecosystem function: need we know more? In press *in* Biodiversity and ecosystem function ed by E-D. Schultze and H. A. Mooney. Springer-Verlag: Heidelberg.

Ehrlich, P. R. and Ehrlich, A. H., 1990. The population explosion. Simon and Schuster: New York.

Ehrlich, P. R. and Ehrlich, A. H., 1991. Healing the planet. Addison-Wesley: New York.

Hartley, W. and Leigh, J., 1979. Plants at risk in Australia. Australian National Parks and Wildlife Service Occasional Paper 3.

Goss, K. and Chatfield, J., 1993. Landcare groups in Western Australia; the role of self help groups in restoring degraded farmland. In press *in* Nature conservation 3: reconstruction of fragmented ecosystems, global and regional perspectives ed by D. A. Saunders, R. J. Hobbs and P. R. Ehrlich. Surrey Beatty & Sons: Chipping Norton, NSW.

Henderson-Sellers, A. and Blong, R., 1989. The greenhouse effect. Living in a warmer Australia. NSW University Press: Kensington, NSW.

Hille, R. J., 1989. Regional scenarios for climate change in WA. Pp. 33–36 *in* Greenhouse 88. Planning for climatic change. Western Australian proceedings ed by G. Lowe. Western Australian Water Resources Council and Environmental Protection Authority: Perth.

Hobbs, R. J. and Atkins, L., 1991. Interactions between annuals and woody perennials in a Western Australian nature reserve. *J. Veg. Sci.* **2:** 643–54.

Hobbs, R. J. and Hopkins, A. J. M., 1991. The role of conservation corridors in a changing climate. Pp. 281–90 *in* Nature conservation 2: The role of corridors ed by D. A. Saunders and R. J. Hobbs. Surrey Beatty & Sons: Chipping Norton, NSW.

Hobbs, R. J. and Saunders, D. A., 1991. Re-integrating fragmented landscapes — a preliminary framework for the Western Australain wheatbelt. *J. Env. Manag.* **33:** 161–67.

Hobbs, R. J. and Saunders, D. A. (eds), 1992. Reintegrating fragmented landscapes: towards sustainable production and nature conservation. Springer-Verlag: New York.

Hobbs, R. J., Saunders, D. A., Lobry de Bruyn, L. and Main, A. R., 1992. Changes in biota. Pp. 65–106 *in* Reintegrating fragmented landscapes: towards sustainable production and nature conservation ed by R. J. Hobbs and D. A. Saunders. Springer-Verlag: New York.

Hopper, S. D., van Leeuwen, S., Brown, A. P. and Patrick, S. J., 1990. Western Australia's endangered flora and other plants under consideration for declaration. Department of Conservation and Land Management: Perth.

Kitchener, D. J., Chapman, A., Dell, J., Muir, B. G. and Palmer, M., 1980. Conservation value for mammals of reserves in the Western Australian wheatbelt. *Biol. Conserv.* **18:** 179–207.

Leigh, J., Boden, R. and Briggs, J., 1984. Extinct and endangered plants of Australia. MacMillan: Melbourne.

Lefroy, E. C., Hobbs, R. J. and Atkins, L. J. 1991. Revegetation guide to the central wheatbelt. Department of Agriculture, CSIRO and Greening Australia: Perth.

Lester, R. T. and Myers, J. R., 1991. Double jeopardy for migrating wildlife. Pp. 119–33 *in* Global climate change and life on earth ed by R. L. Wyman. Routlidge, Chapman and Hall: New York.

Lobry de Bruyn, L. A., 1990. The role of ants and termites in modifying soil properties in naturally vegetated and agricultural environments. Ph.D. Thesis, University of Western Australia.

Lowe, G. (ed), 1989. Greenhouse 88. Planning for climatic change. Western Australian proceedings. Western Australian Water Resources Council and Environmental Protection Authority: Perth.

Main, A. R., 1988. Climatic change and its impact on nature conservation in Australia. Pp. 361–74 *in* Greenhouse. Planning for climate change ed by G. I. Pearman. CSIRO: Melbourne.

Main, A. R., 1993. Restoration ecology and climate change. Pp. 27–32 *in* Nature conservation 3: reconstructing fragmented ecosystems, global and regional perspectives ed by D. A. Saunders, R. J. Hobbs and P. R. Ehrlich. Surrey Beatty & Sons: Chipping Norton, NSW.

Nulsen, R. A., 1992. Changes in soil properties. Pp. 107–45 *in* Reintegrating fragmented landscapes: towards sustainable production and nature conservation ed by R. J. Hobbs and D. A. Saunders. Springer-Verlag: New York.

Pate, J. S. and Beard, J. S. (eds), 1984. Kwongan. Plant life of the sandplain. University of Western Australia Press: Nedlands, WA.

Pearman, G. I. (ed), 1988. Greenhouse. Planning for climate change. CSIRO: Australia.

Peters, R. L., 1991. Consequences of global warming for biological diversity. Pp. 99–118 *in* Global climate change and life on earth ed by R. L. Wyman. Routlidge, Chapman and Hall: New York.

Pittock, A. B., 1988. Actual and anticipated changes in Australia's climate. Pp. 35–51 *in* Greenhouse. Planning for climate change ed by G. I. Pearman. CSIRO: Australia.

Saunders, D. A., 1989. Changes in the avifauna of a region, district and remnant as a result of fragmentation of native vegetation: the wheatbelt of Western Australia. A case study. *Biol. Conserv.* **50:** 99–135.

Saunders, D. A., 1992. The effect of habitat fragmentation on regional bird populations and the need for community involvement in conservation; lessons from the Western Australian wheatbelt. *In* Status and conservation of birds and their habitats in Queensland ed by C. Catterall, P. Driscoll, K. Hulsman and A. Taplin. Queensland Ornithological Society Inc.: Brisbane.

Saunders, D. A. and Curry, P. J., 1990. The impact of agricultural and pastoral industries on birds in the southern half of Western Australia: past, present and future. *Proc. Ecol. Soc. Aust.* **16:** 303–21.

Saunders, D. A., Hopkins, A. J. M. and How, R. A. (eds), 1990. Australian Ecosystems: 200 Years of Utilization, Degradation and Reconstruction. *Proc. Ecol. Soc. Aust.* **16.**

Schneider, S. H., 1989. Global warming. Are we entering the greenhouse century? Sierra Club: San Francisco.

Scougall, A., Majer, J. and Hobbs, R., 1993. Edge effects in grazed and ungrazed Western Australian wheatbelt remnants in relation to ecosystem reconstruction. In press *in* Nature conservation 3: reconstructing fragmented ecosystems, global and regional perspectives ed by D. A. Saunders, R. J. Hobbs and P. R. Ehrlich. Surrey Beatty & Sons: Chipping Norton, NSW.

Taylor, S. G., 1987. Conservation strategies for human-dominated areas: the South Australian example. Pp. 313–22 *in* Nature conservation: the role of remnants of native vegetation ed by D. A. Saunders, G. W. Arnold, A. A. Burbidge and A. J. M. Hopkins. Surrey Beatty & Sons: Chipping Norton, NSW.

van der Sommen, F. J., 1986. Colonization of forests and woodland communities by exotic plants. *In* The ecology of the forests and woodlands of South Australia ed by H. R. Wallace. D. J. Woolman Government Printer: Adelaide.

WCU/UPEP/WWF, 1991. Caring for the earth. A strategy for sustainable living. IUCN: Gland.

Wyman, R. L., 1991. Multiple threats to wildlife. Climate change, acid precipitation and habitat fragmentation. Pp. 134–55 *in* Global climate change and life on earth ed by R. L. Wyman. Routlidge, Chapman and Hall: New York.

Young, M. D., Cocks, K. D. and Humphries, S. E., no date. Australia's environment and its natural resources. CSIRO. Institute of Natural Resources and Environment: Dickson, ACT.

PLATE 2

ABOVE: Ant activity in a relatively intact area of salmon gum *(Eucalyptus salmonophloia)* woodland. Soil fauna play an important role in modifying soil properties and water infiltration. (Photo: R. J. Hobbs).

BELOW: The quokka *Setonix brachyurus* has had a major impact on ecosystem structure and function on Rottnest Island by grazing regenerating tree species and maintaining post-fire vegetation as a low heath rather than allowing development to woodland. (Photo: R. J. Hobbs).

CHAPTER 5

The role of diversity in ecosystem function: an overview

A. R. Main[1]

This chapter argues that biodiversity provides a pool of organisms from which substitutes for system functions can be drawn when conditions change and role dominance alters. The functions discussed are architecture, structure or habitat, carbon and nitrogen fixation, nutrient recycling and retention and interactions which prevent one or a few species dominating the resources of the system. As species richness declines the risk that substitutes will be absent rises as does likelihood of the collapse of the system. The loss or decline of mistletoe from small reserves is used to illustrate the significant changes in species and habitat richness that follow, yet the perception of these might be masked by an assessment that the system was being maintained because the dominant tree was regenerating readily.

Introduction

The retention of representative areas, the preservation of rare and endangered species and maintenance of biodiversity have been accepted as goals of conservation. However, despite the best efforts of management it is likely that, for reasons not amenable to correction by managers, e.g., small reserves in a fragmented landscape or management practices on land adjacent to natural areas, species will be lost from land under their management. The prospect raises the question of the likely significance or consequences of the loss of species from natural assemblages. The simplest manifestation may be merely that fewer species are present; however, in the most extreme case there is the possibility that interactions between species will be disrupted leading to the collapse of the ecosystem. A judgment of where, between these two extremes, the outcome will lie appears to require an intimate

[1]Senior Honorary Research Fellow, Zoology Department, University of Western Australia, Nedlands, Western Australia, Australia 6009.

Pages 77–93 *in* BIODIVERSITY OF MEDITERRANEAN ECOSYSTEMS IN AUSTRALIA ed by Richard J. Hobbs. Surrey Beatty & Sons, 1992.

knowledge of the life histories of all species and their interactions. This is an unachievable goal. This chapter seeks a solution by considering species as they perform system functions rather than solely as entities in taxonomic compilation reflecting species richness.

Concept development

This section addresses the question of what constitutes ecosystem function and how species contribute to it.

When light, water, space, and nutrients are available in a suitable physical environment, photosynthetic organisms can fix carbon. Along with nitrogen provided by symbiotic and free-living nitrogen-fixing organisms, this carbon allows them to produce carbohydrates, amino acids and proteins. These are in turn resources for herbivores, predators, parasites, bacteria and other disease-producing organisms. The net result of the foregoing interactions is that organisms do not live as long or grow as vigorously as they might otherwise do. Following death the remains become resources for other organisms which decompose the tissues and so make nutrients available to bacteria and plants. Well-defined ecosystems are those assemblages which, unlike agricultural or horticultural systems, appear to be self-maintaining with respect to resource requirements. The processes through which this is achieved can be considered as ecosystem functions. These functions result from the activities of representatives of each of the five kingdoms of organisms (Bacteria; Protista; Fungi; Plants; and Animals). Viewed in this light the significance of diversity emerges at a more synthetic level than is usually considered. Not only are resources made available by the actions of individuals of species of this vast array of organisms but by their actions they also prevent the domination of resource use by one or few species. This is a consequence of what Paine (1980) following MacArthur (1972), refers to as the effect of strongly interacting species, defined as follows: "A consumer will be a strong interactor if, in its absence, pronounced changes ensue". Thus strongly interacting species can only be detected in nature following experimental manipulation, but their effects are well recognized and applied in biological control of pests in agriculture and horticulture. Many strongly interacting species are small or minute arthropods, often hymenopterans, but mites, bacterial and viral infections have also been used to depress numbers of pest species. The physical insignificance of many species giving effective biological control suggests that scoring species richness of flowering plants or higher vertebrates may give an inadequate indication of the diversity of the important functional components which maintain the species richness of the system and ultimately lead to the re-use of resources.

Historical

Studies of biodiversity were stimulated by the hypothesis of MacArthur (1955) which postulated that stability in an ecosystem was directly related to the number of pathways in it through which energy could pass. Pathways can be considered in several senses; the first acknowledges that all organisms use energy and so a taxonomic listing of the composition or species present in a system is a measure of the number of pathways present, e.g., studies of food-webs (Pimm 1982). However, not all species do the same things nor are those that do similar things equally efficient at performing under the same or different environmental conditions (Paine 1992; Lawton 1992). A second approach is to consider themes such as energy, nutrient and water fluxes; this approach is useful in considering bulk functions and is amenable to experimental manipulation (Bormann and Likens 1979). Another approach relates to the number of different species that can perform the same functional role, such as nitrogen fixation or recycling, within an ecosystem. This represents the variety of organisms which can substitute for each other, in using energy to do work which maintains the functional integrity of the system.

Taxonomic compilation offers an easily applied method and, if instituted early in monitoring, gives a base-line from which departures in compositon or species abundance can be measured. When further knowledge is to hand it may be possible to identify those groups which are good surrogate measures of functional roles. Meanwhile the loss of a species with little or no apparent effect on the persistence of the system should not be construed as evidence that the species lost was not functionally important to the system. Rather the apparent resilience of the system to the loss may indicate that another species has substituted and now fills the functional role that the lost species formerly performed. If this is true and there are many substitutes present, a great number of species may be lost before the system collapses. It would seem that such a catastrophe can only be avoided by identifying; (a) the ecosystem functions such as the energy consuming pathways of MacArthur (1955) which supply the irreplaceable needs of the system and (b) the array of taxonomic entities which can act as substitutes in performing such essential function. This latter requirement is difficult to satisfy since it depends on the identification of groups which are taxonomically difficult and little researched, so that their biology can be studied in relation to their needs and the conditions under which they contribute most to ecosystem function. Essentially this reduces to a determination of what does best when or where in a fluctuating environment.

The importance of ecosystem functions can be illustrated by considering how simple a system would be if all nutrients or other resources were bound-up in the standing crop of one or a few species; for example, a wheat crop poisoned for both weeds and invertebrate pests. Consumption by herbivores or predators or mortality from other causes normally ensures that this does not happen. The effect of consumption or death is to redistribute the resources elsewhere or make them available to other organisms. In this regard it is of little consequence whether an organism is consumed by a herbivore or predator or attacked by sap or blood-sucking insects or killed by protozoal, bacterial, or viral diseases. Similarly retention of nitrogen within the soil is an important ecosystem function and it is the taxonomic variety of organisms within the soil profile (insect larvae, nematodes, fungi, protozoa and bacteria) which can retain nitrogen and other nutrients within the system under a great variety of environmental conditions which permits functional substitution. A functional approach based on an identification of taxonomic variety with respect to a capacity to substitute roles in performing critical ecosystem functions is a development of the approaches of earlier papers (Main 1981, 1982, 1984, 1987) which is pursued below.

Mediterranean environments

Mediterranean climates are characterized by a cool wet winter and a hot dry summer of varying intensity and duration, occasionally relieved by intense rainfall derived from moist tropical air associated with low pressure troughs along the west coast or tropical cyclonic storms (see Hobbs, Chapter 1). The summer drought may vary in length from year to year depending on the duration of the winter rainy season, which shows much inter-annual variability (see Saunders and Hobbs, Chapter 4). Thus the biota in a mediterranean environment needs to be tolerant of both the extremes and intensity of events likely to be experienced. Furthermore during the hot dry summer the vegetation is very combustible and thus susceptible to naturally occurring fires such as those following lightning strikes. The risk of destruction by fire requires that the biota should not only be able to survive and recover from burning but be able to continue to perform ecosystem functions during the process. A requirement that essential ecosystem functions be performed under all conditions likely to be experienced means that there is scope for taxonomic richness which gives a capacity to substitute one species for another in order to provide critical ecosystem functions over the great variety of environmental conditions which could be experienced. The occurrence at the present time of taxonomically rich assemblages suggests that functional integrity has been achieved and maintained in the past.

Roles

The multiplicity of pathways for energy flow believed by MacArthur (1955) to be important for system stability is provided by individuals of different species which, in satisfying their own biological needs, incidentally fill roles such as abstracting nutrients from soil particles, fixing carbon or nitrogen, providing habitat and resources for other organisms or conserving and sometimes transporting nutrients elsewhere. Within each pathway there is usually more than one species or genotype each being more or less well suited to changed environmental conditions.

What these changed conditions might be can be illustrated by considering nitrogen. Nodulation and nitrogenase activity depend upon the plant species involved as well as on the sensitivity of the symbionts to the nitrate levels and physical conditions in the soil. In mediterranean ecosystems nitrate levels might be expected to be low following fire. Burning stimulates germination and growth of Casuarinaceae, Mimosaceae and Papilionaceae, and at such times symbiotic nitrogen fixation might be expected to be high. However, depending on the tolerances of symbionts and other nitrogen fixing organisms, fixation will fall over time because of the rise in nitrate levels and the short life of many post-fire plants. At this stage the persistence of the system with respect to nitrogen economy will depend upon species other than those fixing nitrogen, whose biology leads to the husbanding of nitrogen within the system, i.e., conserves and recycles it. When the level of nitrate falls below the level which inhibits nodulation or nitrogenase activity, and an energy source is available, fixation can be resumed by either symbiotic or free-living bacteria. The contribution of each to the nitrogen pool will depend upon the potential hosts, species and strains of symbionts present, the presence of free-living organisms capable of fixing nitrogen, as well as the prevailing environmental conditions. At the same time nitrogen can be transported to other locations by solution in water, or following consumption of plants by herbivores or them by predators, or following erosion by water or wind. The net result of these activities is a nitrogen pool held within plants, fungi and soil organisms such as protists, bacteria and invertebrate animals.

This example illustrates the role of diversity in the nitrogen cycle and suggests why species richness is important, but there are many other aspects of ecosystem function, for example the effect of strongly interacting species in preventing one or a few species from dominating resource use. Some important possible functional roles are summarized in Table 1.

Because of the inherent variability of mediterranean environments the components performing each function might be expected to have characteristics which permit the performance of functions under all conditions likely to be experienced. These requirements might be satisfied either by widely tolerant species functioning under an array of conditions, or by species functioning only under a narrow environmental range, or by those with long life cycles but functioning only under favourable conditions which recur frequently but which have a short duration at each recurrence. When history and environment have selected for widely tolerant species then such an ecosystem will have few species capable of filling each functional role, e.g., under conditions of annual or regular burning when only organisms capable of withstanding extremes of light, heat and water availability will persist. Systems with the second or third set of characteristics should have species-rich assemblages and many distinct growth and activity periods, thus providing ecosystem functions under a wide array of environmental conditions (Table 2).

Table 1. Ecosystem processes arranged to emphasize complementary functions.

Process	Function
Photosynthesis	Carbon fixation provides the energy base and the material for the architecture of the ecosystem
Nutrient supply	
nitrogen	Fixation by free living and symbiotic bacteria provides nitrogen for protein synthesis
minerals	Abstraction from soil by plants, mycorrhizae and micro-organisms satisfies physiological needs of biota
transport	Redistribution by herbivores, predators, dust and water avoids local accumulation of nutrients
Recycling of resources	Death and decomposition release nutrients and space for re-use
Maintenance of diversity	Interactions between species prevent one or a few species from dominating resource use

Table 2. Generalized characteristics of potential substitutes, capable of supplying the same needs and so maintaining a functional ecosystem.

Needs	Substitutes
Carbon fixation	The genera and species of green plants with different physiological tolerances and carbon fixing abilities
Nitrogen fixation	Species and strains of nitrogen fixing organisms, free-living bacteria, cyanobacteria
Resource recycling	Mortality induced by herbivory, predation, metazoan parasites, protozoa, bacterial and viral diseases
Retention of diversity	Any non fatal harvesting by herbivores or predators, metazoan parasitism, or chronic protozoan, bacterial, or viral disease which have in common the fact that they reduce the capacity of their host populations to exclude other species from resources

Substitution

In the foregoing the emphasis has been on the possibility that a different species can fill the same functional role as another when environmental conditions change. The extent to which this is true would allay the concern expressed in the introduction that functions may be lost from a system because of local or global extinction of species.

A view of what might be expected following loss of species from Australian mediterranean ecosystems is obscured by:

(a) the presence of exotics such as foxes, rabbits and the fungal pathogen *Phytophthora* which have no strongly interacting species regulating their numbers.

(b) the continued presence of a varied set of species-rich ecosystems despite the loss during the recent geological past of many species (Balme *et al.* 1978; Lundelius 1983; Murray 1984; Merrilees 1984).

(c) further losses of biota since European settlement

(d) fragmentation of the natural environment following agricultural settlement thus restricting, in space and time, the responses of the biota to environmental variations.

(e) lack of knowledge of roles of species in ecosystems, their biology, life histories, and life expectancy.

(f) the lags within the system, related to longevity of individuals, which may obscure or delay the expression of failure of recruitment and the ultimate loss of a species (Frank 1968).

A complete documentation of the changes that are occurring in the biota is desirable but not feasible. Nevertheless it may be possible to select and monitor a group or groups from which to gauge changes. The general importance of the soil (Dindall 1990; Edwards *et al.* 1988) suggests it as a site with many possibilities for selecting a group for monitoring.

Macrofungi as a group contains members which are parasites, saprophytes and mycorrhizal symbionts. They display a wide range of seasonal growth patterns and show different abundances following disturbance. These attributes suggest that fungi, as a group with diverse biologies, may provide a useful surrogate measure of what is happening in the soil and the rooting systems of the asociated plants. Other groups which are more difficult to monitor but which might provide insights into functions performed by soil dwelling organisms are; soil amoebae, as predators of soil bacteria; nematodes as root parasites and hence potential strongly interacting species with respect to plants, whose roots they attack; and bacteria which provide the diet of predatory species.

The effect of losses

The emphasis in this chapter has been on the functional consequences of loss of species from ecosystems. The concept of function has been interpreted broadly to include mortality, which in effect recycles spatial resources, and regeneration which maintains an ecosystem as a dynamic entity. The central theme of the chapter is that a system rich in species will have a potential pool from which substitutes able to fill essential ecosystem functions under a range of environmental conditions can be found. When substitution occurs after loss of species the system may maintain appearances and functions for a variable but potentially considerable period of time. This will depend on how successful the substitution has been, whether environmental change has affected the functional efficiency of the replacement species, and how the resources formerly made available by the lost species are husbanded.

While a number of mammal and bird species have been lost recently from the Australian mediterranean environmments as a consequence of settlement and clearing of native vegetation (see Saunders and Hobbs, Chapter 4), a more serious and continuing loss seems likely in the future as exotic weeds, *Phytophthora,* foxes, rabbits and introduced rodents continue to have an effect on the biota persisting in natural areas.

Species have been lost following environmental changes in the recent geological past and recently following land settlement and the introduction of exotics. Consideration of the outcome of these events suggests that persistence of ecosystems may be affected by the causes of the extinction, the size of the pool from which substitutes may be found, and the space available for range changes. Extinction may finally occur as a result of small population size and demographic stochasticity, but broadly it can be considered under three headings, namely effects of: environmental change, settlement, and exotics; these are discussed in detail below.

Environmental change

The fossil record suggests that from late Pleistocene time through the whole of the Holocene environmental change (Hope 1984) and extinctions have been occurring, so that plant (Churchill 1968; Martin 1973; Kershaw 1981), and animal (Lundelius 1983; Murray 1984; Merilees 1984), distributions have expanded and contracted as the climate has oscillated from wet to dry periods. The present fragmented distribution of some moist habitat plants and invertebrates suggests that drying and fragmentation of distributions was continuing at the time of settlement. Nevertheless the continued presence of species-rich assemblages indicates that under the conditions prevailing ecosystems continued to function, presumably following a series of substitutions as species became extinct. In reality the situation is more complicated. The present composition of the biota gives few clues as to how the Australian ecosystems were reorganized as a consequence of the extinctions, especially following the loss of the megafauna which must have led to significant changes in both their appearance and function.

The megafauna were large to very large animals ranging up to 1 500 kg in weight and a height of two to three metres (Murray 1984), and included large macropods in the genera *Macropus, Protemnodon, Sthenurus* and *Procoptodon;* large quadrupeds such as the monotreme *Zaglossus* (30 kg); the diprotodontids *Palorchestes, Zygomaturus* and *Diprotodon;* koalas, *Phascolarctos;* and wombats, *Phascolonus.* Emus were more diverse and their size range greater than in present-day members of the species (Rich and Van Tets 1984). The large extinct mirihung birds ranged in size from about the size of emus to the largest bird known to have existed. The last of these *(Genyornis newtoni)* survived until at least 26 000 years ago (Rich 1979; Rich and Van Tets 1984). Reconstructions and maps of fossil distribution are given in

Murray (1984), Horton (1984), and Rich and Van Tets (1984). Anatomical reconstructions suggest that the feeding habits of the megafauna ranged from tall browsers to cursorial grazers; some may have lived in flocks or herds in moist woodlands while others occupied drier more open areas. Many possibly had trunk-like appendages and browsed on shrubs or trees. The heavy claws and nasal bones of others suggest a soil-disturbing, pig or tapir-like, feeding mode (Murray 1984). Predators such as *Thylacinus* and *Thylacoleo* (75–100 kg) appear to have been widespread.

Fossil assemblages represent accidents of preservation and not the numerical abundance of species in an environment and give no indication of what was present in places where fossilization was not possible; nevertheless it is clear that in Australia, including that area now subject to a mediterranean climate, there was a fauna capable of browsing and grazing above and below-ground parts of plants of all ecosystems. Such a comprehensive pattern of disturbance ranging from canopy destruction to soil disturbance must have afforded many opportunities for maintaining a diverse flora adapted to regenerate in disturbed patches of various sizes. Even more significant was the potential for browsers of all sizes to affect the structural composition of any post-disturbance (including post-fire) regeneration that occurred.

Notwithstanding the losses referred to above the pollen record indicates that most of the plant species of the biotic environments created by the megafauna persisted after its extinction. Few sites record Late Pleistocene plant communities that could have given rise to the present pattern (Hope 1984). This discrepancy is likely to be related to the way in which loss of the megafauna changed the competitive relationships between plant species. It follows from the foregoing and Hope's (1984) discussion of the climate that, while plant species may have persisted since the Pleistocene, substitution has taken place to yield ecosystems with a different composition and hence identity. The argument suggests that once strongly interacting species became extinct, and as climate changed, the surviving species reorganized into new assemblages and ecosystems within which interactions and feedback loops developed which tended to restabilize the system.

Undoubtedly this is what is to be expected in the future, but fragmentation of the natural environment has restricted the ability of both plants and animals to be shuffled over the landscape as in the Late Pleistocene and Holocene. Substitution may still be possible but its scope in any locality is very limited and, as potential substitutes are locally lost, radical changes in function seem more likely. Consequently, attempts at preservation or retention of assemblages as they are at present is likely to be unachievable.

Settlement

The general disturbance following pastoral settlement and clearing for agriculture has certainly led to the local extinction of plants and animals that formerly occupied what are now pastures or grain fields. Some local relicts persist in areas of remnant vegetation on private land or in nature reserves. The effect of this fragmentation is that the biota is no longer able to respond as it formerly did when deleterious conditions arose, by changing range distributions. The current restriction of many species to remnants of limited size means that they will have to "tough out", by means of physiological or behavioural responses, any environmental changes rather than alter distributions or withdraw to refuges from which populations could later expand (Churchill 1968). Refuges from drought for native biota were exploited early in the pastoral phase of settlement and are no longer available as natural habitat. These limitations are likely to lead to a rapid increase in extinctions in the future. Should the species lost in any new wave of extinctions have performed significant ecosystem functions, then the persistence of the system will depend on substitutes being found to provide these functions. However, the possibility that successful substitution can be made becomes less likely as the pool of possible replacements becomes smaller, or to put it another way as the number of extinctions increases. The foregoing reasoning suggests that ecosystems may collapse suddenly only when no substitutes are available. A more usual scenario might be the gradual loss of species, reduction in species richness, change in dominant species and appearance until the ecosystem lacks any close resemblance to its original state. It is only when these changes are rapid that ordinary observers will take note and express concern.

Exotics

Trampling of litter and soil by hoofed exotics has destroyed the litter layer, broken surface mats of lichens and cyanobacteria, and impaired both the permeabiality and water holding capacity of the soil. As a consequence the soil fauna is reduced in variety and some surviving species are restricted to minor inaccesible sites. The presence of exotic herbivores has inevitably led to the local elimination of plants because of grazing pressure, and predators such as foxes have similarly caused extinction of animal species.

The likelihood that a species pool from which functional substitutes can be drawn will be present after exotics have established is vanishing. The net result will be that ecosystems will lose their characteristic appearance because functions such as carbon fixation, structure and architecture, species interactions and nutrient husbandry become impaired. Moreover, since exotics lack strongly interacting species to influence their abundance they are likely to increase by exploiting resources until these are monopolized, as in the case of plants, or destroyed, as with animals suffering from intense predation.

In the presence of *Phytophthora* spp. it is impossible to maintain the composition and structure of vegetation except in areas unsuited to the pathogen or in habitats occupied by resistant species. In the presence of rabbits and foxes system functions can be maintained only where there are plant species (*Gastrolobium* spp.) containing the poison monosodium fluoroacetate, 1080, to which endemic species are tolerant or resistant (Twigg and King 1991).

The persistence of ecosystems in the presence of exotics depends on the ability of the system to supply substitutes for the roles and functions listed in Table 1. The way these functions are maintained in the face of changing conditions will vary (see Table 2). The question of substitution is central to the ability to retain ecosystem function and hence the general appearance of an assemblage. Conceptually substitution can be at the level of replacing species or genotypes within species or populations. These replacements can have environmental tolerances which are either broad generalists, or narrow specialists. The ability of specialists to function only within a narrow environmental range requires, as a corollary, that they be numerous and replace each other as environmental conditions alter. In practice conditions are likely to be more diffuse than the foregoing might suggest.

Notwithstanding the different levels at which substitution can take place it is those ecosystems losing species from the critcal pathways with fewest potential substitutes that are at greatest risk of collapse. However, it is not apparent from a list of species present what might be the function at risk, e.g., structural architecture, nutrients or regulation of species. These questions can only be resolved by using a comparative methodology, by following natural changes or by deliberate manipulation of a situation. In essence the problem becomes one of biology and ultimately population biology in the numerical, demographic and genetic senses. However, this solution is largely academic while exotics are having such a large impact on the local persistence of species. The continued presence of exotics will, by attrition of the numbers of endemic species, ultimately lead to the collapse of function in all ecosystems affected by them because the capacity to substitute will be lost.

Case Study

The foregoing sections set out the consequences of loss of species; in summary the argument suggests that should substitutes be present it may be difficult to detect an impaired function. Moreover, identification of lost functions may be obscured by our perception of a single trait as an indicator of what constitutes an intact system. An example from the woodlands of mediterranean Western Australia illustrates such a case.

These woodlands were dominated by either wandoo *(Eucalyptus wandoo, E. capillosa)* or york gum *(E. loxophleba)*, depending on soil types with an understorey of jam *(Acacia acuminata)* common to both. Clearing for agriculture has reduced the area of these woodlands. Burrows *et al.* (1991) have shown *E. wandoo* to regenerate best on valley floors after fire, but all species regenerate in disturbed areas and stands of all ages can be found in even small remnants. The architectural structure of these woodlands provides the habitat for mistletoes (*Amyema* spp.) which can be common and abundant on wandoo. As with all parasites these probably reduce the vigour and life expectancy of their host by killing boughs and so make light and space available to other plants (see Fig. 1). Barlow (1981) argues that the similarity of leaves of mistletoe and its hosts' leaves is an evolved mimicry which provides protective concealment from vertebrate herbivores. Local anectdotal evidence suggests that, in the past, brush-tailed possums *(Trichosurus vulpecula)* were possibly a major herbivore of mistletoes. Thus there is a likelihood of a strong interaction between the mistletoe and the folivores which ate them along with eucalypt leaves and other foliage. This interaction continued long after settlement and even in the presence of foxes while the poison plants (*Gastrolobium* spp.) containing monosodium fluoroacetate were abundant. In the absence of the poison plants foxes soon eliminated possums. It might be expected that, in the absence of the herbivore, mistletoe would become abundant and kill host trees as in South Australia (Lewis 1991). This has not happened.

Fig. 1. Wandoo *(Eucalyptus wandoo)*, Chester Pass, Stirling Range National Park, showing dead limbs in infected plant and uninfected adjacent trees. (Photograph B. Y. Main).

Mistletoes are widespread but in the agricultural areas their occurrence is localized, contagious and obviously related to a source of infection (Lamont 1985). Infection is commonly mediated by the mistletoe bird *(Dicaeum hirundinaceum)* which eats the fruits and disperses the plant by defecating the sticky seeds on to boughs of trees and shrubs where they germinate and establish. The rate of passage is short, 20–40 minutes (Barlow 1981), or slightly longer as determined in feeding trials by Keast (1958). Barlow (1981) believes dispersal is most likely over short distances with long range dispersal only if seeds stick to feathers.

The mistletoe bird is nomadic but tends to breed where there is a source of fruit. This period of residence might be expected to enhance re-infection of old hosts and infection of new ones. Re-infection is a prerequisite for the persistence of mistletoe because individual plants have a life expectancy of less than 15 years (Barlow 1981). Reid (1987) argues that mistletoe in Australia evolved in relation to autochthonous birds such as honeyeaters (Meliphagidae) and not the mistletoe bird which arrived in the Pleistocene. Honeyeaters which might be considered as alternatives for the mistletoe bird are affected by the same attributes of vegetation remnants as is the mistletoe bird. Moreover even if they are common they are unlikely to be more efficient at distributing seed. Lamont and Southall (1982) showed a difference in density of mistletoes along a road verge relative to the adjacent nature reserve and interpreted this as visitation pressure by bird dispersers using the road verge as a corridor. Prior to settlement woodlands were extensive. At present they are small and by chance may lack sufficient mistletoe to attract seed vectors regularly either in a nomadic or breeding phase. From the foregoing one might predict a high rate of extinction of mistletoe in remnants and a low chance of re-infection should an adjacent infected remnant be distant from that suffering local extinction. The observed contagious distribution of mistletoe and a low rate of infection of regenerating trees suggests that this may be so.

The larvae of two butterflies, *Ogyris* sp. (Lycaenidae), and *Delias aganippe* (Pieridae), feed on mistletoe. Formerly they were widespread and sometimes common but this is no longer so — a phenomenon no doubt associated with the decline in abundance of mistletoe. They formerly visited flowering shrubs and herbs, so, to the extent that they were pollen vectors and hence contributed to gene flow, this function has been lost.

The significance of this tale is twofold; various functions have been lost; and the ready regeneration of components, especially wandoo, which maintains the architectural structure of the system gives an illusory impression that all is well and that these woodlands are not much affected by fragmentation following clearing for agriculture. A more appropriate generalization may be that the consequences of the loss of functions provided by a diverse fauna may not be readily apparent.

Conclusion

In this chapter I have argued that the role of diversity in an ecosystem is to provide a pool of organisms from which functional substitutes can be found following the loss of organisms that formerly filled that role. The important functions are to: fix carbon and nitrogen; provide an architectural structure or habitat; interact with other species so that one or several do not dominate the use of resources; and recycle resources so that the net loss from the system is minimized. While substitutes are present the loss of diversity may not be manifest as loss of function. Systems only become at risk when substitutes cannot be found and even so the loss may not be detected because, as shown by an example, a very conspicuous functional attribute may obscure the loss.

The large number of extinctions and range contractions already experienced in the fragmented natural environments of the mediterranean climatic zone of Australia has massively reduced the capacity of ecosystems to engage in species substitution to fill essential roles, as probably successfully happened following the extinction of the megafauna. It follows that prudence is called for in conservation practice because, should substitution be no longer possible but needed in the changed environment of a greenhouse world, the systems may be near to the point where further losses will lead to the loss of any resemblance to the ecosystems now recognized.

References

Balme, J., Merilees, D. and Porter, J. K., 1978. Late Quaternary mammal remains, spanning about 30 000 years, from excavations in Devil's Lair, Western Australia. *J. Roy. Soc. West. Aust.* **61**: 33–65.

Barlow, B. A., 1981. The loranthaceous mistletoes in Australia. Pp. 555–74 *in* Ecological biogeography of Australia ed by A. Keast. Junk: The Hague.

Bormann, F. H. and Likens, G. E., 1979. Pattern and process in a forested ecosystem: disturbance, development and the steady state based on the Hubbard Brook ecosystem study. Springer-Verlag: New York.

Burrows, N., Gardner, C., Ward, B. and Robinson, A., 1990. Regeneration of *Eucalyptus wandoo* following fire. *Aust. For.* **53**: 248–58.

Churchill, D., 1968. The distribution and prehistory of *Eucalyptus diversicolor* F. Muell., *E. marginata* Donn ex Sm. and *E. calophylla* R. Br. in relation to rainfall. *Aust. J. Bot.* **16**: 125–51.

Dindall, D. L. (ed), 1990. Soil biology guide. John Wiley and Son: New York.

Edwards, C. A., Stinner, B. R., Stinner, D. and Rabatin, S., 1988. Biological interactions in soil. Elsevier: Amsterdam.

Frank, P. W., 1968. Life histories and community stability. *Ecology* **49**: 355–56.

Hope, G., 1984. Australian environmental change: timing, directions, magnitudes, and rates. Pp. 681–90 *in* Quaternary extinctions: a prehistoric revolution ed by P. S. Martin and R. G. Klein. University of Arizona Press: Tuscon.

Horton, D. R., 1984. Red kangaroos: last of the Australian megafauna. Pp. 639–68 *in* Quarternary extinctions: a prehistoric revolution ed by P. S. Martin and R. G. Klein. University of Arizona Press: Tuscon.

Keast, A., 1958. The influence of ecology on variation in the mistletoe-bird *(Dicaeum hirundinaceum) Emu* **58**: 195–206.

Kershaw, A. P., 1981. Quaternary vegetation and environments. Pp. 83–101 *in* Ecological biogeography of Australia ed by A. Keast. Junk: The Hague.

Lamont, B., 1985. Host distribution, potassium content, water relations and control of two co-occurring mistletoe species. *J. Roy. Soc. West. Aust.* **68**: 21–25.

Lamont, B. B. and Southall, K., 1982. Biology of the mistletoe *Amyema preissii* on road verges and undisturbed vegetation. *Search* **13**: 87–88.

Lawton, J. H., 1992. Feeble links in food webs. *Nature* **355**: 19–20.

Lewis, S. A., 1991. The conservation and management of roadside vegetation in South Australia. Pp. 313–18 *in* Nature Conservation 2: the role of corridors ed by D. A. Saunders and R. J. Hobbs. Surrey Beatty & Sons: Chipping Norton, NSW.

Lundelius, E. L., 1983. Climatic implications of Late Pleistocene and Holocene faunal associations in Australia. *Alcheringa* **7**: 125–49.

MacArthur, R., 1955. Fluctuations of animal populations and a measure of community stability. *Ecology* **36**: 533–36.

MacArthur, R., 1972. Strong or weak interactions? *Trans. Connecticut Acad. Arts Sci.* **44**: 179–88.

Martin, H., 1973. Palynology and historical ecology of some cave excavations in the Australian Nullarbor. *Aust. J. Bot.* **21**: 283–316.

Main, A. R., 1981. Ecosystem theory and management. *J. Roy. Soc. West. Aust.* **64**: 1–4.

Main, A. R., 1982. Rare species: precious or dross? Pp. 163–74 *in* Species at risk: research in Australia ed by R. H. Groves and W. D. L. Ride. Australian Academy of Science: Canberra.

Main, A. R., 1984. Rare species: problems of conservation. *Search* **15**: 94–97.

Main, A. R., 1987. Management of remnants of native vegetation. Pp. 1–13 *in* Nature Conservation: the role of remnants of native vegetation ed by D. A. Saunders, G. W. Arnold, A. A. Burbidge and A. J. M. Hopkins. Surrey Beatty & Sons: Chipping Norton, NSW.

Merrilees, D., 1984. Comings and goings of Late Quarternary mammals in extreme southwestern Australia. Pp. 629–38 *in* Quaternary extinctions: a prehistoric revolution ed by P. S. Martin and R. G. Klein. The University of Arizona Press: Tuscon.

Murray, P., 1984. Extinctions downunder: a bestiary of extinct Australian Late Pleistocene monotremes and marsupials. Pp. 600–27 *in* Quaternary extinctions: a prehistoric revolution ed by P. S. Martin and R. G. Klein. The University of Arizona Press: Tuscon.

Paine, R. T., 1980. Food webs: linkage interaction, strength and community infrastructure. *J. Anim. Ecol.* **49**: 667–85.

Paine, R. T., 1992. Food web analysis through field measurement of interaction strength. *Nature* **355**: 73–75.

Pimm, S. L., 1982. Food webs. Chapman and Hall: London.

Reid, N., 1987. The mistletoe bird: coincidence or coevolution. *Emu* **87**: 130–31.

Rich, P. V., 1979. The Dromornithidae, a family of large extinct ground birds endemic to Australia. *Bureau of Mineral Resources Bulletin* 184.

Rich, P. V. and Van Tets, G., 1984. What fossil birds contribute towards an understanding of origin and development of the Australian avifauna. Pp. 421–46 *in* Vertebrate zoogeography and evolution in Australia ed by M. Archer and G. Clayton. Hesperian Press: Carlisle, Western Australia.

Twigg, L. E. and King, D. R., 1991. The impact of fluoroacetate-bearing vegetation on native Australian fauna: a review. *Oikos* **61**: 412–30.

PLATE 3

LEFT: Carnaby's cockatoo *Calyptorhynchus funereus latirostris* is a species native to southwestern Australia which has declined dramatically in abundance and range over the past few decades. This species may have important effects on regeneration of *Banksia* species through its predation on parasitic wood borers (see Chapter 6). (Photo: G. Chapman).

BELOW: Individual plant species, such as this *Banksia prionotes,* may play important roles in maintaining whole elements of the fauna. (Photo: R. J. Hobbs).

CHAPTER 6

Functional interactions within plants — the contribution of keystone and other species to biological diversity

Byron B. Lamont[1]

The way plants function is considered a component of biodiversity. Many examples exist in mediterranean Australia of plant-abiotic, plant-plant, plant-animal and plant-microbe interactions which have major effects on productivity, population sizes, species composition and ecosystem function. Symbiotic microbes are most likely to contribute to the highest order functional pathways, ultimately controlling the way the ecosystem functions and thus biodiversity. The major constraints on productivity, apart from extraneous influences associated with technocratic humans, are poor soils, summer drought and bushfires. Rare species contribute more to biodiversity than to ecosystem functioning. The simplest balanced ecosystem conceivable contrasts greatly with species composition of such floristically-rich systems as the heathlands (kwongan). The keystone concept shows some promise as a means of valuing species for their contribution to ecosystem function. Further analysis indicates that there is in fact a hierarchy of relationships between species in terms of (a) the number of functional pathways to which they contribute, directly or indirectly, and (b) their quantitative contribution to these pathways. Issues for future research include identification of ecosystems with greatest or novel biological diversity and keystone species which contribute most to functional pathways.

Introduction

This chapter explores the contribution of plants to ecosystem functioning, as a component of biodiversity, in the context of mediterranean Australia. My premise is that the contribution of a species to biodiversity depends on the number of functional pathways with which it interacts. Examples are provided of plant-abiotic, plant-plant, plant-animal and plant-microbe interactions

---

[1]School of Environmental Biology, Curtin University of Technology, PO Box U1987, Perth, Western Australia, Australia 6001.

Pages 95–127 *in* BIODIVERSITY OF MEDITERRANEAN ECOSYSTEMS IN AUSTRALIA ed by Richard J. Hobbs. Surrey Beatty & Sons, 1992.

and permutations between them. High biodiversity can be equated with intricate sharing of the resources by biologically-diverse components, while differences in biodiversity may reside with contrasts in the pattern of resource use. This is preceded by sections on definitions and threats to diversity. The discussion addresses such questions as "Can rare species contribute to ecosystem function? What are the constraints on biodiversity? How many species are required for a functional ecosystem? Are there better ways of valuing species? How can the functional role of a plant species be determined?", using examples from the text to reach conclusions.

Definitions

Biodiversity is a conceptually-rich term which unfortunately is often considered merely synonymous with species diversity. I consider *biodiversity* to be shorthand for "biological diversity", as accepted officially in Australia (Commonwealth of Australia 1992), and not "biotic diversity" which implies only the taxonomic component. This opens the way for considering all disciplines associated with the abundance and variation in organisms — ecology, genetics, physiology, anatomy, systematics, conservation, and at all levels — cell, individual, population, community, regional. I define *function* as a biotic process that involves or facilitates transfer of energy or matter (see Hobbs, Chapter 1 for other aspects). A complete description of a particular function requires stating what is transferred, the genotypes/species involved, the transfer pathway between them or with their abiotic environment, and the extent of transfer under prevailing conditions. Transfers are often mediated by organisms other than the species of prime interest — e.g., pollen transport by animals, nutrient uptake via symbiotic microbes. In this context, a *guild* is a group of functionally-related species, i.e., they participate in the same pathway but not necessarily to the same extent.

It may be argued that the contribution of a species to biodiversity depends on the number of species with which it interacts, directly or indirectly. This is a continuum, but analyses could be helped by grouping species into those essential for the continued presence of just one other species in the ecosystem, essential for two or more (n–1) species, and essential for all *(n)* species, i.e., maintenance of the ecosystem. The last category has been described as a *keystone species* (Paine 1969). The Macquarie dictionary (1981) defines keystone as "1. the wedge-shaped piece at the summit of an arch, regarded as holding the other pieces in place. 2. something on which associated things depend". I therefore define a *first-order* (arch) keystone species as essential for the continued presence of one other species within an ecosystem. A first-order *keystone guild* becomes a group of functionally-related species, individually or collectively responsible for the continued presence of another species. A *second-order* (vault) keystone species has an essential (usually the same) function for a few species. (A vault keystone is illustrated in Corbeil and Manser 1988). A *third-order* (cathedral) keystone species has a prime role in functioning of the ecosystem, consistent with the original concept of keystone species (Payne 1969). Its multiplier effect is so great that in its absence the ecosystem would collapse. This could be by interacting with most species, directly or indirectly, or through using most resources (the dominant species). It could simply be a keystone for another species essential for the functioning of the ecosystem. A third-order keystone guild is a group of functionally-related species that has a prime role in functioning of the ecosystem.

Threats to plant biodiversity

There are many threats to the maintenance of biodiversity in mediterranean Australia. Some of these influences have general effects while others target functional groups. Wholesale clearing for agriculture, and to a lesser extent for towns, roads and mining, has eliminated vast tracts of vegetation in southern Australia (Saunders and Hobbs, Chapter 4). Invasion by exotic grasses and forbs is a major hindrance to re-establishment of indigenous plant species after clearing or fire (Hobbs and Atkins 1991). Inappropriate fire intervals are most likely to lead to local extinction of fire-killed species (Burgman and Lamont 1992). Acid rain is not a problem here but superphosphatic fertilizers are a major pollutant of soils and waterways — some farmers use "super" to kill the native heath before sowing to pasture. Introduced pathogens and herbivores, like the weeds, have reached epidemic proportions. In particular, *Phytophthora cinnamomi* is a generalist root pathogen to which the major family Proteaceae is particularly susceptible (Wills 1992). Other lethal *Phytophthora* species, canker and rust diseases are becoming widespread. Rabbits, now immune to *Myxomatosis,* are on the increase and favour seedlings and graminoids. While bedevilled by the coarseness of the data base, one disturbing scenario for southwestern Australia, in response to rising carbon dioxide levels, is an escalation in the trend for increasing aridity and higher temperatures (Pearman 1988;

Saunders and Hobbs, Chapter 4). With lack of opportunities for genetic change or even migration, local and global extinctions of plant species may follow this trend. There will be some buffering capacity because of the long history of long, hot, dry summers here but these are already the limiting part of their life cycle (Lamont *et al.* 1991a; Lamont *et al.* 1992).

All the above lead to a direct loss of biodiversity. But there are equally important *indirect* effects too, especially if the impacts are mainly felt by keystone groups. This includes the detrimental, and potentially escalating, effect of the loss of beneficial organisms. Such losses set in train a linear sequence of events that may lead to ecosystem collapse, rather than the cyclic events that maintain ecosystem functions.

Interactions affecting ecosystem function

I now work through contrasting examples of interactions affecting the functioning of plant communities in southwestern Australia. Components of biodiversity examined include species composition and numbers, population sizes (equitability), species contribution to the landscape (physiognomy, including visual impact), taxonomic significance, comparative morphology, functional pathways and whole-plant responses to the major environmental constraints. Plants interact with their abiotic environment such that their population sizes continually fluctuate and may even be reduced to zero under particularly severe conditions. Plant species interact with each other, the outcome again affecting population sizes and even presence/absence of species in the community. Plant-animal interactions are particularly important in ecosystem function, where they usually operate at the level of guilds. Plant interactions with microbes greatly add to the functional complexity of the ecosystem — in their absence, most higher plants would be eliminated gradually. Plant-abiotic-animal, plant-animal-animal and plant-microbe-animal interactions also occur, with their indirect effects perhaps contributing most to biodiversity. For example, with two species there are two (direct) interactions, while if they interact with a third there are potentially six interactions with three indirect.

Plant-abiotic interactions

These are fairly well studied, especially in relation to the role of soil type in species distribution, and fire and drought on species demography. However, use of the data for predictive purposes via stochastic modelling and sensitivity assessment (population viability analyses) is still in its infancy (Lavorel and Noble, Chapter 8).

Species-rich heathlands

The regions of greatest species richness are centred on the so-called northern and southern sandplains — scrub and mallee-heath (kwongan) on deep sands and shallow laterites. George *et al.* (1979) identified up to 48 vascular species in 1 m^2 plots for lateritic heath at Mt Lesueur which may well be a record for any vegetation type. These can be considered some of the least fertile soils in the world (Lamont 1992a) with moderate water availability (Lamont and Bergl 1991). Since each soil type tends to have its own characteristic flora, Hopper (1979) attributed high landscape species richness to the mosaic of soil types. Others have emphasized ecotonal soils as the sites of greatest species richness at a finer scale (Hopkins and Hnatiuk 1981; Brown and Hopkins 1983; Froend 1987).

Floristic richness of the heathlands (kwongan) is very high at a local scale. It also functions differently from other vegetation types — resources are shared more equitably among the components (Lamont *et al.* 1977); a tree stratum is lacking; annuals are rare; much effort is expended into flowering (the famous spring wildflowers) yet seed set is usually less than 1 per cent (Lamont *et al.* 1985a); most species have strongly-developed supplementary mechanisms for nutrient uptake (Lamont 1982, 1984); the degree of sclerophylly (leaf density and thickness) is exceptionally high (Witkowski and Lamont 1991, Lamont 1992a) and the leaves are correspondingly long-lived (Witkowski *et al.* 1992); canopy-reaching fires are certain within the lifespan of all species, with most resprouting after fire and storing seeds in the soil or (especially) in the plant canopy (Lamont *et al.* 1991b).

On the other hand, there appears to be much duplication in function. First and second-order keystone species exist, but third-order species are unlikely. This can be attributed to two sources: firstly, richness is at the level of species, rather than genera or families. Of the 89 families in kwongan, five account for 56 per cent of species, while the 10 major genera (2.3% of the total) contain 731 species, 29.2 per cent of the total (Lamont *et al.* 1984). Secondly, the combination of constraints is so restrictive that there are few biologically realistic solutions available for dealing with them (Lamont and Bergl 1991). Rather than competition avoidance (which requires alternative functioning) it is more likely to be low level competition tolerance (the same functions, at least as adults).

Drought and fire

The nature of the weather pattern following fire is crucial for successful recruitment of many species into the new generation (Lamont and Witkowski 1992). *Banksia cuneata,* with less than 600 known adults, is the rarest and most threatened species of this important genus in mediterranean Australia. Its special contributions to biological diversity include close morphological links with the large genus, *Dryandra,* endemic to southwestern Australia; among the smallest flower heads and least number of follicles per cone in the genus; a tendency to form localized thickets affecting the distribution of other species; a major nectar source for insects and birds in spring; highly combustible and thus potentially a major propagator of fires. It possibly ranks with *B. prionotes* as a keystone species for nectarivorous birds in the Western Australian wheatbelt (see Lambeck, Chapter 7). Following an experimental fire in an isolated road verge population, the mean of 780 germinants per (fire-killed) adult was reduced to almost zero following an exceptionally dry summer (Lamont *et al.* 1991a). Placement in local depressions and artificial watering resulted in 90 per cent survival the following year. Since the year was 35 per cent wetter, survival among the controls (ridge, no extra water) reached 40 per cent. We have modelled the results to show that without fires the population is likely to become extinct eventually and with fires it will become extinct sooner, especially at short intervals (Burgman and Lamont 1992). Were this species widespread, it would have a major impact on the functioning of the landscape. Currently, its demise will only have a very localized impact. Functionally, its place could be taken by *Dryandra sessilis,* and perhaps *Adenanthos cygnorum,* although they never co-occur with *B. cuneata,* so that the main losses would be taxonomic, genetic and aesthetic.

Not all species respond to drought in the same way. Some seedlings use water sparingly and are more likely to survive the first summer after fire. These are usually the needle or linear-leaved rather than broad-leaved species (Enright and Lamont 1989, 1992; Lamont *et al.* 1992). Depending on the weather pattern in the first few years after fire, this could affect relative population sizes and hence the way the community functions with respect to water use. In contrast, the larger-seeded species (and seedlings from larger seeds of the same species) have fast initial growth rates, reaching and sustaining contact with stored soil water better than small-seeded, slow-growing species. *Banksia speciosa* (large broad leaves) fits the first category while the co-occurring *B. pulchella* (small needle leaves) and *B. coccinea* (small broad leaves) fit the second category (Lamont and Witkowski 1992). Following a wet summer, seedling/parent ratios were not significantly different between the three species. Following the next (exceptionally hot, dry) summer, *B. speciosa* was still above self-replacement levels, but the two small species were well below. Under the current fire regime, the scenario is for an ecosystem whose functioning is increasingly dominated by the vigorous, highly fecund *B. speciosa* (Table 1). Certainly, *B. speciosa* is a third-order keystone species, but the other species may serve as lower-order keystones for resident pollinators as well as contributing quite differently to the structure of the vegetation. Altering the fire regime appears the only option for decreasing dominance by *B. speciosa* — spring fires at 5–10 year intervals (Table 1), as *B. coccinea* is short-lived and has little canopy seed storage (Witkowski *et al.* 1991), while seedling establishment of *B. speciosa* is enhanced preferentially by autumn fires. At this age interval however, seed production is so low, especially for *B. pulchella,* that the stochasticity of seedling recruitment means local extinction could be an unfortunate outcome.

Summer heat

Less is known about the effect of summer heat. Some species have remarkably high heat tolerances (up to 60°C in *Hakea preissii*), increased by lower water availability and higher leaf nitrogen

Table 1. Ratio of mean number of genets (adults, viable seeds, seedlings) per 400 m^2 for three co-occurring banksias with *B. pulchella* set at 1. Modified from Lamont and Witkowski (1992).

Genet type	*B. coccinea*	:	*B. speciosa*	Total genets
Adults at 5 years	1.3		21.0	107
Seed store at 5 years	28.2		64.2	77
Seed store at 10 years	39.5		89.9	531
Seed store at 15 years	9.2		84.1	2 290
Adults at 21 years	0.7		21.0	105
Seed store at 21 years	0.7		79.1	13 107
Seed release after spring fire	1.9		136.2	8 090
Seed release after autumn fire	0.7		79.4	12 154
Seedlings after first summer (mean)	3.8		60.5	373
Seedlings after second summer				
— after spring fire	0.8		39.1	97
— after autumn fire	1.5		214.3	221

(Hellmuth 1971; Zohar *et al.* 1981; Lamont and Kelly 1988; Lamont unpubl.). Thus, not only do different species cease to function metabolically at different temperature extremes, the limit depends on the growing conditions of the habitat. Following two successive days of the highest air temperatures recorded in the Albany region (45–47°C) accompanied by hot winds, we examined the effects on 50 major species over the next few months (Lamont and S. Leighton, unpub.). Most plants of some species, e.g., *Hakea ferruginea,* were killed (non-sprouters, endemic to the area, large leaves, shrubs in the open) while others, e.g., *Banksia attenuata,* showed no effects (resprouters, also present nearer Perth, highly sclerophyllous leaves or leafless, trees or rhizomatous herbs protected from the winds or in shade). This single event, like fire, will have a profound effect on the functioning of the living landscape for years ahead. Unlike regeneration after fire, the opportunities for seedling recruitment are restricted because competition with surviving adults is intense (Lamont *et al.* 1992) and substantial canopy seed release in some species requires fire (Witkowski *et al.* 1991).

Plant-plant interactions

Interplant relations operate at the population and community levels and may be positive (facilitatory) or negative (competitive) in their effects on community function.

Parasitic flowering plants

One of the clearest keystone relationships is that of parasitic flowering plants with their obligate dependence on flowering plant hosts. There are at least 42 parasitic species in mediterranean Western Australia (Lamont 1982) that would be absent if there were no suitable hosts. Host specificity (the extent to which a given host species is a keystone) varies greatly from non-existent (most of the root hemiparasites) to sub-generic (the stem parasites). The mistletoes in particular may act as lower order keystone species for dependent animals (Main, Chapter 5), but they do not act as keystones for other plants — this includes *Nuytsia floribunda* which is prominent in many vegetation types and has extraordinary biological attributes (Fineran and Hocking 1983; Lamont 1985a). Mistletoes in particular alter the functional relationships of the community by diverting a large proportion of the remobilizable nutrients, especially K and P, from the host to other pathways (Lamont 1983), including ultimately to some frugivores and herbivores. Nutrients released from large root parasites accumulate under their canopies and foster growth of nutrient demanding herbs there (J. S. Pate, pers. comm.) as well as altering the pattern of nutrient distribution through the soil.

Interplant competition

Competition for water can be acute among seedlings. In a thinning field trial, not only did the thinned seedlings of two banksias and a hakea grow larger, but survival by the end of the first summer was up to five times greater (Table 2; Lamont *et al.* 1992). This intense competition was evident only in the post-fire litter microsites where most seeds accumulated and not in the surrounding sand patches where nutrient availability was lower and water availability was higher. The type of fire affected the properties of the patch types as well as the patterns of seed distribution and seedling survival. For example, after a patchy (backburn) fire, seedlings of the resprouter, *Banksia attenuata,* increased at the expense of the non-sprouting *B. hookeriana*, but not after an intense wildfire (Table 2). The shifts in relative numbers of these dominant resprouters and non-sprouters following different fires has important implications

Table 2. Effect of patch type following two adjoining fires on seedling distribution and survival. Seedlings were thinned to a third of their initial density in some litter patches. Different superscript letters indicate significant differences based on the SNK multiple range test. Asterisks indicate significant differences between thinned and unthinned patches based on χ^2 tests. Extracted from Lamont *et al.* (1992).

	Wildfire			Backburn		
	Litter		Sand	Litter		Sand
Patch area (% total)	12.6^{d}		87.4^{a}	67.1^{b}		32.9^{c}
Litter concentration (g m^{-2} patch)	193.9^{a}		7.9^{b}	123.5^{a}		31.9^{b}
Nitrogen (mg g^{-1})	3.3^{a}		0.3^{b}	2.1^{a}		0.5^{b}
Phosphorus ($\mu g\ g^{-1}$)	24.7^{a}		1.3^{b}	5.0^{b}		1.9^{b}
Initial seedlings (no. m^{-2} patch)	284.1^{a}		6.9^{c}	66.3^{b}		12.7^{c}
Post-summer seedlings (% survival):	Control	Thinned		Control	Thinned	
Banksia attenuata (flat-leaf, resprouter)	37	74*	68	22	50*	62
B. hookeriana (flat-leaf, non-sprouter)	33	58*	80	7	35*	64
Hakea polyanthema (needle, non-sprouter)	81	100*	92	78	93	96
Shoot dry mass (g):						
B. attenuata	1.0^{b}	1.3^{b}	1.7^{ab}	1.3^{b}	1.7^{ab}	2.0^{a}
B. hookeriana	1.2^{ab}	1.1^{ab}	1.6^{a}	0.8^{b}	1.2^{ab}	1.5^{ab}
H. polyanthema	1.5^{b}	2.0^{ab}	2.5^{a}	1.6^{b}	2.1^{ab}	2.3^{a}

for community function. In contrast, survival of the needle-leaved hakea was little affected by seedling density or patch type. *B. hookeriana* is part of a keystone guild of banksias (Lamont and Bergl 1991) although it may be a keystone in its own right because of its role as a major nectar source for bird pollinators and its winter flowering.

The distribution of some species may be limited by edaphic-competition interactions. Unlike three other co-dominant *Banksia* species in the landscape, *B. hookeriana* is restricted to the crests and slopes of deep sand dunes (Lamont *et al.* 1989). The initial interpretation is that the local distribution of this species is under strong edaphic control (see, for example, Groves and Hobbs, Chapter 3). However, we obtained equal germination and post-summer survival of *B. hookeriana* in monoculture at three contrasting locations along a topographic gradient. When the level of interspecific competition was raised, by adding seeds of the other three banksias, germination was unaffected but only those *B. hookeriana* in the deep sand site survived. *B. hookeriana* at the sites where it was eventually excluded was under much greater water stress than where it persisted. Thus, the presence of poor competitors in marginal habitats (and hence the functioning of the ecosystem) is affected by the level of competition, i.e., identity and abundance of other species there (Fig. 1). This is consistent with the Chesson and Huntly (1988) negatively-synergistic model which predicts such species will not co-occur with strongly competitive species in fluctuating environments. Thus, *B. hookeriana* is restricted to locations where soils are conducive to growth and interspecific competition between seedlings is minimal (Lamont *et al.* 1992; Enright and Lamont 1992).

Plant-animal interactions

We now consider plant-animal relationships from the point of view of the plant. There are few cases of first-order keystone animals, with some notable exceptions in the field of pollination.

Pollination

Within the terrestrial orchids, there are many species dependent on a single insect pollinator (Stoutamine 1983). For example, all seven species of *Drakaea* have their own (male) wasp species (family Thynnidae) — each mimicking the associated female wasp in morphology, colour and pheromone production (Stoutamire 1975). Attempts to carry away the mimic (articulated labellum)

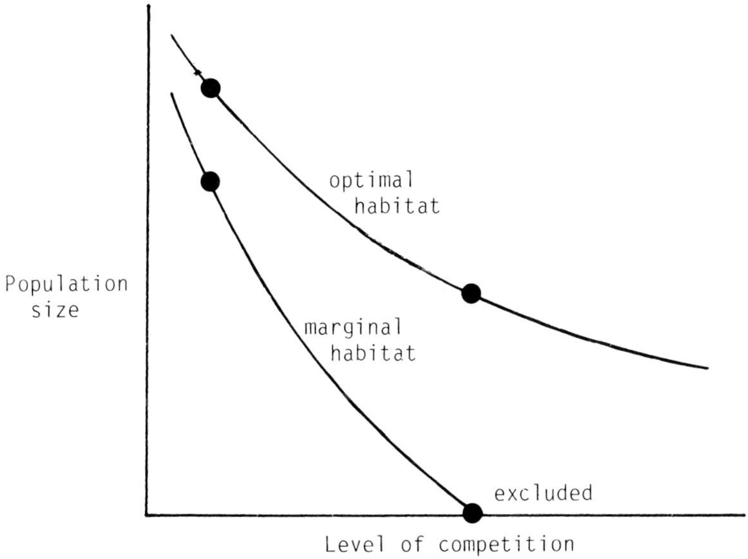

Fig. 1. Interpretation of the survival levels for *Banksia hookeriana* along a topographic gradient (Lamont *et al.* 1989) in terms of the negative synergistic model of Chesson and Huntly (1988). This class of species is a poor competitor whose presence in the ecosystem is dependent on the level of competition from other co-occurring species and likelihood of marginal conditions for survival. This produces a mosaic of *B. hookeriana* populations across the landscape. This in turn affects the activities of animals (pollinators, granivores) dependent on this species for their presence in the landscape. Spots represent the actual data points.

results in the male brushing against the viscid disc of the pollinia or receptive stigma. Loss of any wasp species would result in the demise of the orchid as well. The loss of any animal-mimic pair would be a blow to biodiversity but have negligible impact on functioning of the ecosystem at any scale other than that of the individual plant.

Each of two closely-related verticordias in the Myrtaceae (*Verticordia nitens, V. aurea)* has its own bee pollinator (Houston *et al.* 1992; unpubl.). The pollen is secreted in an oily fluid into a helmet above the anther. This is terminated by a spout from which the bee laps the mixture, squeezing the helmet between its mandibles at the same time (Fig. 2). Negligible seeds are set if the bee is absent. *V. nitens* is the major understorey species in parts of banksia woodland north of Perth and a number of invertebrate herbivores and carnivores appear to depend on it (Houston unpubl.). As it is unlikely to lead to community collapse in its absence, *V. nitens* can be considered a second-order keystone with respect to the animals. The bee will be a first-order keystone with respect to *V. nitens.* Other verticordias are pollinated by guilds of insects (Lamont 1985b). Many species in the mediterranean Australian flora are pollinated by guilds of unrelated animals — insects, birds and small mammals (Collins and Rebelo 1987). Such faunistically-rich guilds make the likelihood of effective pollination particularly robust in the face of the loss of any one animal species.

Banksia tricuspis has a special place in the Mt Lesueur scrub-heath, 200 km north of Perth. It is a long-lived small tree considered a relict among banksias, and gazetted as rare, but is locally dominant and conspicuous because of its tendency to form emergent patches at high points in the landscape (George 1981). The habitat of *B. tricuspis* is so restricted and specific (fissures in lateritized sandstone in exposed locations with a 500 mm annual rainfall) that it is doubtful that there exists any functionally-equivalent species to replace it. Five or more honeyeater species (family Maliphagidae) and the honey possum *(Tarsipes rostratus)* are the significant pollinators of *B. tricuspis* (S. van Leeuwen, unpubl.). Since this species has a high level of outcrossing, loss of the poorly mobile honey possum would have little impact on its long-term survival (but see Cunningham 1991). The effect of loss of honeyeaters would vary from negligible (one species) to major

Fig. 2. The sole (keystone) pollinator, *Euryglossa morrisoni* (Hymenoptera), of the locally-dominant understorey shrub, *Verticordia nitens. (a) (opposite page)* male bee lapping oily pollen from the highly modified anther. *(b) (top)* female bee reaching down to the nectary, receiving pollen on its ventral surfaces as it does so. In the absence of this bee few seed are set. Flowers about 10 mm wide. *(c) (bottom)* Stand of *Verticordia nitens* in banksia woodland. Photos: T. Houston (a and b) and B. Lamont (c).

(all species) in the long term. The effect of the absence of the pollinator guild (i.e., negligible seed production) would not be detected for hundreds of years, as few seeds released after fire yield recruits to the resprouting population and fire tolerance of seedlings is not achieved for at least 20 years after fire (Lamont and van Leeuwen 1988). Thus, populations are almost stable independent of fire regime or seed availability.

Herbivory

Parrots have a surprisingly important role in enhancing seed set in *B. tricuspis*. Most (65%) flower heads are destroyed by wood-boring grubs of the moth *Arthrophora* sp., Tortricidae (Fig. 3; S. van Leeuwen, unpubl.; Lamont 1992b). The black cockatoo *(Calyptorhynchus funereus latirostris),* and to a much lesser extent corellas *(Cacatua sanguinea)* feed on the grubs. In consuming 84 per cent of these insects they only make a "mistake" 6 per cent of the time, i.e., breaking open extra heads that do not contain the grubs, while missing only 16 per cent of those that contain them. The cockatoo thus acts as an agent of biological control, presumably increasing the fitness of this banksia by enabling it to set extra seeds in subsequent crops through reduced abundance of the adult insects. The cockatoo can be regarded as a first-order keystone species for as long as the *Arthrophora* moth has a potentially large role in reducing seed set, i.e., its effect is largely indirect. The moth can be considered a first-order keystone because of its direct negative effect on reproduction of *B. tricuspis*. Again, the absence of omnivorous birds on adult *B. tricuspis* populations would not be evident for hundreds of years; although the stands would function differently immediately — population explosion of the moths, reduced food for pollinators, negligible seed set or release, no post-fire seedlings.

The woolly bush, *Adenanthos cygnorum* (Proteaceae), is one of the species which gives character to the Swan Coastal Plain. It may form grey thickets after the banksia woodland is removed for landclearing purposes or after fire. It is the major successional species following extensive soil disturbance on the way to restoration of the banksia woodland. It has a major role in maintaining resident nectar-feeding birds as it flowers from spring to autumn (Newland and Wooller 1985). Many insects feed from the extra-floral nectaries on some of the leaves while the fruits may be eaten by birds and rodents (Lamont 1989). Many ant species consume the aril (elaiosome). *A. cygnorum* is a major litter producer and, once dead, may be consumed by termites before the next fire.

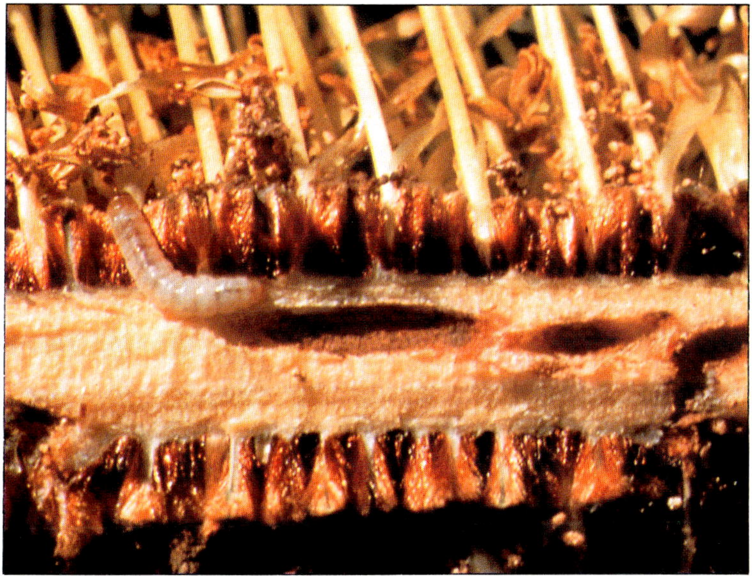

Fig. 3. Flower head of *Banksia tricuspis* cut longitudinally to show grub, *Arthrophora* sp., that tunnels the rachis and prevents fruit set. It is controlled by the black cockatoo *(Calyptorhynchus funereus latirostris)* which is able to detect heads containing grubs and avoids heads without them. The bird thus acts as a facultative first-order keystone species for the banksia. Grub about 10 mm long. Photo: S. van Leeuwen.

Although the obligatory nature of many of these relations has not been established, it is clear that this species can be regarded as a third-order keystone (Fig. 4). Were the ants not to carry the fruits from the plant canopies to their nests then the fruits would be consumed by vertebrate granivores (Lamont and Grey 1984). Burial enables the fruits to remain dormant until the soil is disturbed or a fire occurs — often after above-ground evidence of the presence of the species is lost. The ant guild is therefore a first-order keystone for the woolly bush. While not usually attractive to herbivores, there is a specialist phytophagous moth, *Xylorycta* sp. (Xyloryctidae), which webs together the terminal leaves and destroys the stem apices. It is preyed upon by a parasitoid wasp, *Campoletis* sp. (Ichneumonidae), which also feeds from the extra-floral nectaries. While seven of 10 xyloryctid pupae that we hatched yielded *Campoletis* wasps (Lamont 1989), the effect of removing this parasitoid from the ecosystem on the long-term persistence of *A. cygnorum* cannot be guessed at without experimental studies.

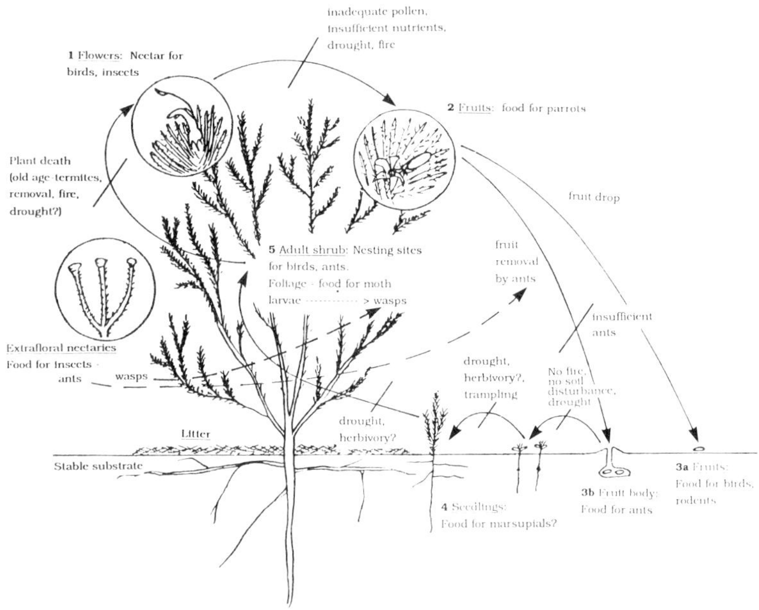

Fig. 4. Major events in the life cycle of *Adenanthos cygnorum* (numbers refer to the sequence), highlighting the interactions between various animals and the plant. Lines through the arrows indicate reasons that particular function might fail. Ants and nectarivorous birds comprise keystone guilds for the plant. From Lamont (1989).

Plant-microbe interactions

Micro-organisms may have profound direct negative (pathogenic) or positive (symbiotic) effects on plants, quite apart from their indirect (saprophytic, autotrophic) effects. They also contribute enormously to biodiversity in their own right. Microbes add greatly to the functional complexity of ecosystems. The activity of free-living microbes is enhanced by the presence of growing roots (the rhizosphere effect). Some of these decompose organic matter while others release nutrients from inorganic sources, thus facilitating nutrient uptake by the roots.

Proteoid roots

Over 500 species of Proteaceae in southwestern Australia produce proteoid roots (Lamont 1982). The stimulus for proteoid root production is provided by free-living soil bacteria, possibly *Bacillus* spp. (Lamont 1982). These dense clusters of rootlets are essential for uptake of poorly-soluble ions (P, Fe, Mn) in impoverished soils (Grierson and Attiwell 1989). However, they are optional structures and may be absent altogether in fertile soils (Lamont 1972). The proteoid root mat enmeshed among the decomposing litter also has a major effect on soil stability and structure (Low and Lamont 1990) as well as affecting the passage of water through the soil. The contribution of proteoid roots to the plant's root system thus greatly affects the way it functions. For the many vegetation types dominated by Proteaceae it also affects the way the community functions. These proteoid-root bearing species form a third-order keystone guild ultimately dependent on a keystone bacterial guild that is always abundant in the organically-rich surface soil.

Root nodules

Other microbes affect the host's morphology and functioning by invading the tissues. These include N_2-fixing bacteria, *Bradyrhizobium* spp., which stimulate root nodule production among all peas (Fabaceae) and wattles (Mimosaceae); N_2-fixing actinomycetes, *Frankia* spp., which stimulate formation of balls of nodules with root extensions in Casuarinaceae; N_2-fixing cyanobacteria, *Nostoc* spp., which modify coralloid root morphology in *Macrozamia riedlei* (Cycadaceae) (Lamont 1982, 1984). Monk *et al.* (1981) estimated that N_2-fixation on the Swan Coastal Plain accounted for 8 per cent of the N content of *Acacia pulchella* (Fig. 5) in its first year rising to 68 per cent in the fourth year, and that 56 per cent of N returned to the soil was derived from N_2-fixation. In the absence of N in the medium, nodulated *Casuarina cunninghamiana* accounted for all plant N (Rodriguez-Barrucco *et al.* 1970). At 3.8 ppm N in the medium, the plants were 3.9 times larger and fixed N contributed only 9 per cent of plant N. Thus, these symbiotic procaryotes may make a major contribution to the way the plant functions, depending on soil fertility in particular (Fig. 6).

It seems likely, in the case of plants that are normally heavily-nodulated, that absence of the microsymbionts would lead to demise of the seedlings before reaching maturity. There are probably only a few microbial species within each host-specific group as outlined above comprising the keystone guild for symbiotic N_2-fixation. While the native legumes can accept a wide range of rhizobia, usually only one "cross-inoculation" group having a wide host tolerance is common in the acid soils of southern Australia (Lamont 1982). If major N_2-fixers, such as *Acacia pulchella*,

Fig. 5. Acacia pulchella in banksia woodland on the Swan Coastal Plain.

were to be eliminated through loss of the appropriate microsymbiont, then N availability in the ecosystem might fall below critical levels for support of the dominants. System collapse would take hundreds of years because most trees and major shrubs are long-lived resprouters. An exception would be communities dominated by N_2-fixers, such as *Casuarina* and *Acacia* thickets, which do not resprout after fire but rely on seedlings for regeneration. These scenarios could be tested by removal of legumes from the ecosystem (Hobbs, Chapter 11). The essential role of bacteria for the legumes could not be explored as it is impossible to envisage conditions that would result in selective elimination of N_2-fixing microsymbionts from the surface soil but not other microbes.

Since many plant species normally occupy the N_2-fixing guild within a community, it is instructive to consider their ability to replace each other functionally should, for some reason, one or more of the species be eliminated. The first point to be made is that the legume-rhizobium combination is at least one to two orders of magnitude more efficient than the non-legume combinations on the basis of the C_2H_2-reduction technique (Lamont 1982, 1984). Secondly, rates of fixation may vary considerably between congenerics at the same site — for example, Shea and Kitt (1976) recorded 1.7 μmol C_2H_2 reduced/g fresh nodule/h for

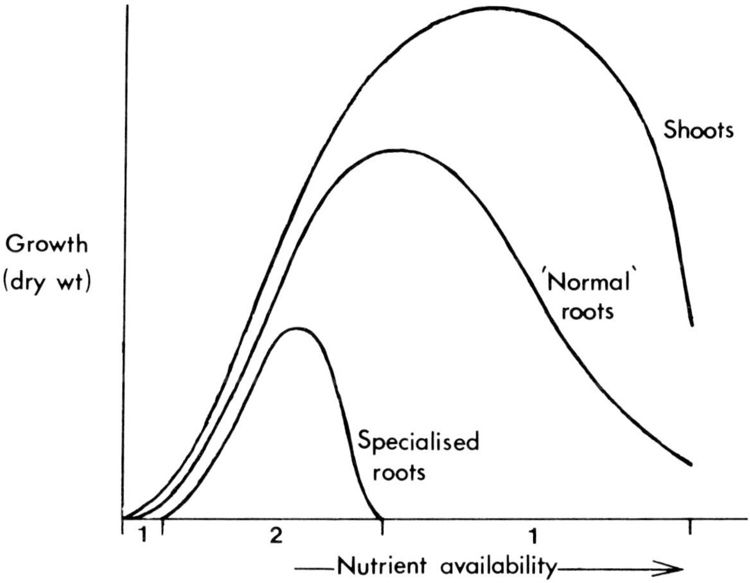

Fig. 6. Generalized response of specialized (nodules, mycorrhizae) and "normal" roots and shoots to increasing soil nutrient availability. 1 = monomorphic root system (no specialized roots), 2 = dimorphic root system. The varying relations between the three organs have major impacts on the way the plant functions, as well as the ecosystem where these are major contributors to biomass. From Lamont (1982).

Acacia extensa, 2.0 units for *A. myrtifolia* and 3.8 units for *A. pulchella* on laterite. When differences in nodule mass per plant and (actual or potential) density of plants are taken into account, this could lead to such major differences in N contribution to the community that the long-term outcomes for ecosystem maintenance could be quite different. In seeking surrogates, it is not just a question of whether they participate in critical pathways but the extent of *quantitative participation* in these pathways.

Mycorrhizae

Hundreds of fungal species probably form symbiotic associations with thousands of plant species in mediterranean Australia (Lamont 1982, 1984). Like root nodules, they modify root morphology and function to varying degrees. A few species in the Endogonaceae (Abbott and Robson 1977) infect the surface rootlets of species in most vascular families here. In exchange for carbohydrates/organic acids from the host, the vesicular-arbuscular

(VA) hyphae increase general mineral ion uptake and enhance the shoot concentrations of P in particular and also Zn, K, S and Ca. An increase in concentration accompanying extra growth indicates that these nutrients were limiting growth of the non-mycorrhizal plant, so the special advantage of mycorrhizae may be the correction of nutrient inbalances, especially of P. Although VA-infected species of genera represented in mediterranean Australia have shown marked growth responses over the controls (Lamont 1982), there have been no local studies. While there can be no doubt that VA mycorrhizae greatly affect the way a plant functions, that this is essential for its continued presence in the community remains untested. The relationship is further complicated by studies which show rhizosphere bacteria may increase VA-infection and host growth markedly (Azcón *et al.* 1978). It is worth noting that VA-infection is considered a pre-requisite for nodule formation in legumes (e.g., Mosse *et al.* 1976). These indirect effects make the interactions below ground as equally complex as those above ground.

Many major trees and shrubs in mediterranean Australia are ectomycorrhizal, including *Eucalyptus diversicolor,* one of the world's largest trees. The hyphae form a dense sheath around the rootlets and cause considerable modifications to their anatomy and morphology (Chilvers and Pryor 1965). Many genera of macro-fungi in the Basidiomycetes form ectomycorrhizae (Gardner and Malajczuk 1988) and a given root system may even possess a mosaic of different species (Chilvers 1973). There are many reports of growth responses to sheathing mycorrhizae among eucalypts in particular (Lamont 1982; Lamont *et al.* 1985a). These mycorrhizae or cultures of their fungi can enhance absorption of P and NH_4 in particular plus NO_3, K, Ca, Cl, SO_4, Mg, Fe, Cu and Zn (Bowen 1973). There is growing evidence of the capacity of ectomycorrhizae to mineralize normally unavailable sources of P via phosphatase secretion, including organic P from litter before it has entered the soil (Herrara *et al.* 1978; MacFall *et al.* 1991). Since their energy requirements are met by the host, ectomycor-rhizal fungi probably reduce litter decomposition by outcompeting saprophytes for access to mineral nutrients (Gadgil and Gadgil 1971). Aluminium, known for its inhibition of phosphatase activity, is a major component of the fossil soils of mediterranean Australia. There is some evidence that organic acids secreted by the fungi chelate and thus detoxify Al released during solubilization (Lamont 1982). They also deter root pathogens and protect the apices from adverse soil conditions such as low pH (Trappe 1977).

Clearly, ectomycorrhizal fungi have a major effect on the way the plant functions (a first-order keystone guild of numerous species), and where ectomycorrhizal species control vegetation structure, as they usually do, a major impact on the way the community functions (a third-order keystone guild of scores of species).

As with VA mycorrhizae, ectomycorrhizae are redundant for nutrient uptake in fertile soils and different species and sources are not equally effective on the same host in poor soils. For example, Table 3 shows that the presence of mycorrhizal fungi makes no difference to growth of *E. diversicolor* seedlings when 16 mg P is added per kg of soil, but it does at lower levels (Bougher *et al.* 1990). At low levels of added P, the effect on plant mass of the three fungal species used varied by up to 27 times, probably due to different effects on root growth as well as variation in ability to enhance P uptake. P uptake is still enhanced at the higher levels but it is no longer limiting growth. Clearly, the *identity* of ectomycorrhizal species dominating the root system could make a profound effect on early rates of growth and probably survival of this keystone eucalypt and hence play a central role in controlling the functioning of the ecosystem. In practice, this might prove unlikely as many different ectomycorrhizal fungi are usually available simultaneously for inoculation (Gardner and Malajczuk 1988). In attempting to manipulate forest ecosystems for commercial purposes it raises the prospect of increasing productivity by fostering certain fungi but this will reduce diversity at the taxonomic and functional levels.

Table 3. Growth and root P content of karri (*Eucalyptus diversicolor*) seedlings inoculated with different ectomycorrhizal fungi in soil at three levels of applied P. Different letters indicate significant differences between vertical means by Duncan's multiple range test. Selected results from Bougher *et al.* (1990).

Fungal inoculum	mg P applied kg^{-1} soil		
	0	4	16
(a) Plant dry mass (g)			
Descolea maculata	0.22^{a}	0.81^{b}	2.95^{a}
Laccaria laccata	0.16^{ab}	1.90^{a}	3.46^{a}
Pisolithus tinctorius	0.07^{b}	0.07^{c}	2.66^{a}
Control	0.09^{b}	0.09^{c}	2.46^{a}
(b) mg P content/g fine root			
Descolea maculata	0.55^{b}	1.50^{a}	2.62^{a}
Laccaria laccata	0.74^{a}	1.49^{a}	2.14^{a}
Control	0.38^{c}	0.45^{b}	1.42^{b}

Plant-microbe-animal interactions

Gastrolobium bilobum forms dense thickets in the southern jarrah forest where it is a major N_2-fixer, soil stabilizer, cover for vertebrates and food for 1080-resistant animals (Lamont 1991). When grown in its normal soil autoclaved, seedlings were over nine times larger in the presence of ectomycorrhizae than in their absence (Lamont *et al.* 1985b). In the absence of the symbiotic fungi, the *G. bilobum* seedlings were far too weak to have survived in the field. The underground sporocarps produced by most of these fungi are the favoured food of a small marsupial, *Bettongia penicillata* (Christensen 1980). Most spores in the faecal pellets belong to the major mycorrhizal genus *Mesophellia*. The spores are much more likely to germinate after passing through the marsupial's body than fresh spores (Lamont *et al.* 1985). Not only does consumption of the sporocarps increase after fire, *Bettongia* may travel up to 3 km overnight as it moves from one patch to another. Our work has shown that the pellets will increase the numbers of mycorrhizal types on a given root system from an average of three to five, in a range of soils and fire histories (Lamont 1992b).

G. bilobum can be regarded as a third-order keystone in certain forest communities. It takes part in a complex set of reciprocal relationships, from indirect to obligate (Fig. 7). The species is dependent on an ectomycorrhizal guild of which some *(Mesophellia)* species are more abundant or have greater affinity

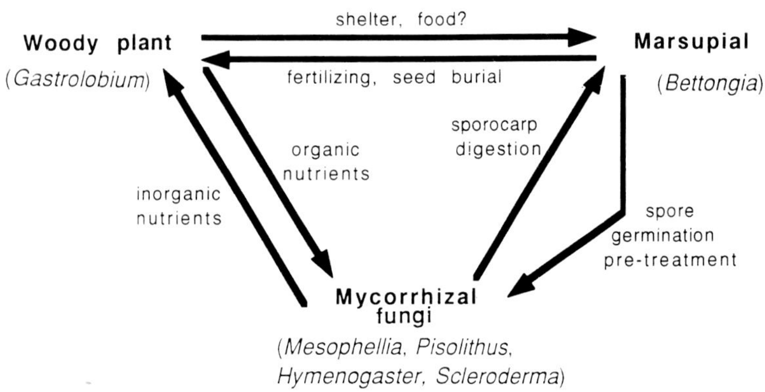

Fig. 7. Interrelations between the woody plant, *Gastrolobium bilobum*, ecto-mycorrhizal fungi and the small marsupial, *Bettongia penicillata*. All components are dependent on each other, highlighting the potential complexity of functional relations within ecosystems.

for its root systems than others (C. Ralph, unpubl.). Loss of fungal species would therefore have negligible to major impact on the presence of the plant species in the long term (the rate depending on incidence of fire and fungal identity). The loss of the marsupial may not affect population viability for hundreds of years because fires may only partially sterilize the surface soil (Malajczuk and Hingston 1981). In addition, a few other mammals *(Isoodon, Rattus)* may devour sporocarps opportunistically. The place of *Bettongia* as a keystone in the functioning of the ecosystem remains a tantalizing possibility. Since this once abundant marsupial is now absent from much of the jarrah forest, this hypothesis could be tested by returning it to selected areas and monitoring any vegetation changes.

Discussion

Biodiversity, ecosystem function and keystones

While preparing this chapter I came to appreciate the need to define biodiversity as widely as possible. Only then do we understand the magnitude of the task of conserving biodiversity or ramifications of manipulating ecosystems. To species composition should be added the abundance of each species at any given point in time. At the landscape level it should include the pattern of distribution of populations through space. Diversity should refer to the range in morphology and structure of the phenotypes as well as their taxonomy. It is worth noting that most symbiotic microbes have a structural as well as functional effect on the root systems of their hosts. Visual and other sensory impacts of the species go beyond morphology and physiognomy and include prominence in the landscape because of position, synchronized mass flowering and even aromas emitted by leaves. Taxonomy should include species' evolutionary status and identification of ancestral and derived characters. Functional relations should be considered a component of biodiversity. Indirect relations between organisms may have as profound an effect on their functioning as direct effects. Witness the role of the black cockatoo in controlling moth predators on flower heads of *Banksia tricuspis.* Functions should go beyond pathways and identify the quantitative contribution to transfers of matter by each species under stated conditions. For example, there may be orders of magnitude difference between the effect of different symbiotic microbes on nutrient uptake under nutrient-poor conditions while all may be ineffective in fertile soils.

Against the subtleties of high biodiversity are the crudities of the forces currently reducing it. Wholesale loss of highly complex ecosystems from landclearing by bulldozers for agriculture, urbanization and mining continues unabated in mediterranean Australia (Saunders and Hobbs, Chapter 4). These biological deserts imply new "ecosystems" by definition and may involve huge fluxes of matter through components introduced to the system by technocratic humans (Hobbs, Chapter 11) The ecosystem functions are reduced to the simplest pathways and offer few insights into the relationship between functional complexity and biodiversity. In contrast, examination of the way plant species interact with the physical environment with which they evolved and with other indigenous plants, animals and microbes has revealed much about the way floristically-rich ecosystems function.

The keystone species concept (Paine 1969) appears useful for identifying the functionally important species in an ecosystem. In practice, it is difficult to explore empirically (see "Issues for future research") and most examples remain speculative and anecdotal (Bond 1993). In addition, there is value in distinguishing a hierarchy of functionally important species, varying from those with a minor impact on species composition (first-order keystone) to a major impact on ecosystem functioning (third-order keystone). An example of how these concepts can be applied to an ecosystem is given in Figure 8. Most interacting species can be grouped into keystone guilds, i.e., they contribute to the same functional pathway. However, if one member of the guild is lost long-term changes could still occur because each contributes to a different extent, i.e., the ecosystem ultimately would function differently. Recall the many fold different effects on growth of karri by various species of ectomycorrhizal fungi. It is also doubtful that available surrogates for the black cockatoo and bettong would be equally effective should these animals be lost from the ecosystem. Thus there is a hierarchy of quantitative relationships within functional groups such that only those at the top, individually or collectively, should be treated as keystones.

Rare species and ecosystem function

Can rare species contribute to ecosystem function? Much depends on whether rarity refers to within or between ecosystems. My examples show even gazetted rare species such as *Banksia cuneata* and *B. tricuspis* may be locally prominent (they are rare on a global basis) and contribute much to functioning of their communities as

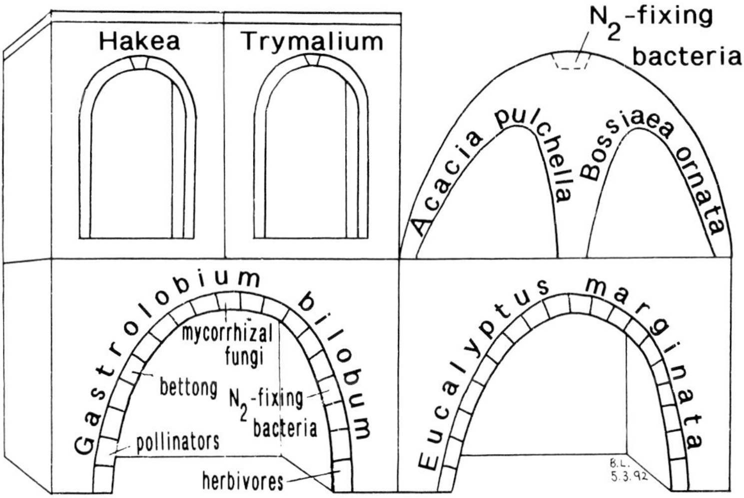

Fig. 8. Metaphor of keystone species in a cathedral (mosque?) using the jarrah forest at Perup as an example. The *Hakea* and *Trymalium* species are not essential to functioning of the ecosystem, so that loss of species associated with maintaining them in the ecosystem has no further effect (first-order keystones). *Acacia pulchella* and *Bossaiea ornata* are members of a group of legumes dependent on a N_2-fixing bacterium or guild of such microbes, whose absence would lead to demise of all legumes in the ecosystem — the bacteria are therefore second-order keystones. *Gastrolobium bilobum* is essential to the functioning of the ecosystem as described in the text (third-order keystone). If it is accepted that the ectomycorrhizal fungi are a keystone guild for *G. bilobum* then their absence would lead to loss of *G. bilobum* and eventually to ecosystem collapse (third-order keystone also).

well as to biodiversity. Unless they are higher order keystones, species contribution to ecosystem functioning is proportional to their direct contribution to resource utilization. Even when species rare in an ecosystem are keystones, they are likely to be lower order e.g., the thynnid pollinators of hammer orchids (both taxa are rare). Demise of both groups would have little impact on functioning of the ecosystem but, because of the unique nature of the relationship, would be a great loss to biological diversity. That is, their contribution to biodiversity far outweighs their contribution to ecosystem functioning. Other rare species in ecosystems are most likely to belong to functional guilds — if they increased the fitness of their associated species their numbers would be expected to increase at the expense of other members, so that if they had recently evolved they would not stay rare for long. While higher order keystones are not necessarily dominants, they are unlikely to be rare either — because of the quantification problem outlined earlier. Determining contribution of a rare species ecosystem function would involve detailed study of its population dynamics and ecophysiology, highlighting inter-taxa relations, rather than removal experiments whose effects would be far too subtle or long term.

Constraints on biodiversity

What are the major constraints governing biodiversity in mediterranean Australia? We considered the generalist impacts of landclearing, vermin and exotic pathogens earlier and noted how they greatly reduced biodiversity and simplified functional pathways. Within ecosystems not subject to these extraneous influences, rainfall, soil type and fire regime appear to be the major resource constraints on biodiversity. Fire regime is considered a resource constraint here as it has a profound effect on the pattern of carbon and mineral turnover and water use. These three factors not only control the presence/absence of species in the ecosystem but also their population sizes. The areas of greatest species diversity (highest species richness and equitability) are those with the most infertile soils (sands and laterites) and moderate rainfall (400–600 mm yr^{-1}) and are fire-prone (non-sprouting components more likely to die from fire than old age). Whether these are also the regions of greatest biodiversity is a moot point — I have argued elsewhere (Lamont 1982, 1984) that their functional complexity below ground matches that above ground. However, it is already clear that biodiversity is not proportional to the size of the limiting resources.

Minimal species for a functional ecosystem

How many species are required for a functional ecosystem? This question is the ultimate in simplification of the issues, and clearly, the starting point must be a primary producer (autotroph). To minimize functional complexity one choice is a free-living cyanobacterium capable of fixing N_2, e.g., *Nostoc*. It would only be seasonally-active and, to avoid desiccation, would need shade for most of the day, say from rock ledges (already not an option for the vast tracts of sands in southern Australia). Because the soils are deficient in most essential nutrients and the cyanobacterium is usable to solubilize or access non-N nutrients in the moister, non-surface part of the soil profile, free-living heterotrophic bacteria or fungi would be required to interact with the autotroph. This leads to *Nostoc*-bearing lichens, e.g., *Leptogium*, as the best formula for stable substrates. This assumes there is a suitable lichen combination available for the different rock types.

The next N_2-fixing symbiosis up the evolutionary tree (and trying to minimize functional complexity) in seeking a colonizer of the sands will be the cycad, *Macrozamia riedlei.* Immediately the ecosystem is complicated many fold: a suitable *Nostoc* species must be available — a fire is usually required to create lesions in the coralloid roots to enable the bacteria to enter; the cycad has a weak root system and must associate with a VA mycorrhizal fungus to enhance P uptake. While it is usually accepted that dioecious species (another component of biodiversity) are wind-pollinated, there is increasing evidence that cycads are pollinated by beetles — I have seen pollen-covered weevils emerging from the cones of *M. riedlei.* This immediately raises the issues of the requirements for the beetle to complete its life cycle and the distribution of the sex forms through the landscape in relation to mobility of the beetle. The seeds are distributed by a wide range of vertebrates, of which the emu is clearly the most effective. Since the animals only feed opportunistically on the seeds (coning is usually stimulated by fire) the normal food and life-cycle requirements of at least the emu would also need to be considered as part of the functioning of the ecosystem. Fires are clearly crucial to the efficient functioning of the cycad — they also stimulate coralloid root growth and N_2-fixation. If their density is low, additional highly combustible plants would be needed in the gaps to ensure propagation of the flames. At the very least (ignoring guilds), the simplest sandy ecosystem would require seven species, yet give the impression of only a monoculture of cycads stretching between lichen-covered rock outcrops.

Valuing species

Are there better ways of valuing species than trying to determine their contribution to ecosystem function? Because the concept of biological diversity can potentially cover all variability within and between species it has much scope for assessing the contribution of phenotypes and genotypes to ecosystems. It thus embraces trophic relations, dominants, evolutionary status and keystones. The functional approach is particularly powerful as it directly addresses the sustainability of the ecosystem via interactions between the components. It provides a basis for predicting the likelihood of system change or collapse when components are modified or lost. This is formalized via keystone species and guilds, but even elaborated upon as I have done here it is limited as it deals with the number of functional pathways rather than the quantification of those pathways. Recently, there has been an attempt to value species for conservation purposes by cladistic analysis (e.g., Erwin 1991). This gives highest value to the most recently-evolved species — as the ones supposedly most able to respond to environmental change. These might well contribute to the unique properties of an ecosystem but must take second place to keystone species which maintain the integrity of the ecosystem in the first place.

Issues for future research

While often a reasonable guess can be made about the contribution of a given species to biodiversity through a study of its population dynamics and ecophysiology (e.g., Lamont and Witkowski 1992), its role is only confirmed following its removal from, or return to, the ecosystem. Pot trials are useful for quantifying the functional relationships between seedlings and potential microsymbionts but give no information on their likely survival in the field — sometimes smaller plants survive better than larger ones (Enright and Lamont 1992). Unfortunately, removal has its own problems — pulling disturbs the root systems of remaining species while severing at the base leaves the root system to contribute organic matter, nutrients and solution channels for growth of other species; should the removed shoots be lifted out of the stand or left to decompose?; it is simply not possible to remove large trees and shrubs without impacting on the rest of the ecosystem while doing so; to attain a new equilibrium may take many generations; it is virtually impossible to delete selected microbes from the soil — most techniques sterilize everything or leave large groups of species unaffected, e.g., antibiotics have no affect on fungi. And yet it is soil microbes that are most likely to be the source of the highest-order keystones and their individual functions may be quite different.

If it is accepted that the priority ecosystems for conservation are those with greatest or novel biodiversity, then all ecosystems need to be re-examined from a biological rather than biotic perspective. The biggest challenge is to determine and quantify the functional pathways between ecosystem components. This will enable identification of the likely keystone functions and species. Experimental deletion (where possible — see above) of likely keystones, at as high an order as possible, should be implemented and the effects monitored. The results should be modelled to enable predictions to be made in the face of future threats to the sustainability of the ecosystem. The modelling should focus on sensitivity analyses — i.e., identifying the components to which the system most responds — and the projected outcome of using surrogates (species with the same functional pathways). Despite high rates of losses in species diversity at present, these studies will not appeal to many because of their long-term nature. Then much thought may need to be given at the outset to the opportunities for short-circuiting the steps and maximizing return for effort (see Lavorel and Noble, Chapter 8; Hobbs, Chapter 11). I hope it is not too late!

Acknowledgements

I thank William Bond for drawing the keystone concept to my attention, Louise Egerton-Warburton for updating me on the mycorrhizal literature and other co-workers over the years for providing me with the knowledge base on which to do justice to this large topic, and Sandra Lavorel, Ray Wills and Richard Hobbs for their useful, if not time-consuming!, comments on the manuscript.

References

Abbott, L. K. and Robson, A. D., 1977. The distribution and abundance of vesicular-arbuscular endophytes in some Western Australian soils. *Aust. J. Bot.* **25**: 515–22.

Azcón, R., Azcón-G. De Aquilar, C. and Barea, J. M., 1978. Effects of plant hormones present in bacterial cultures on the formation and responses of VA endomycorrhiza. *New Phytol.* **80**: 359–64.

Bond, W. J., 1993. Keystone species. In press *in* Ecosystem function of biodiversity ed by E.-D. Schulze and H. A. Mooney. Springer-Verlag: Heidelberg.

Bougher, N. L., Grove, T. S. and Malajuzuk, N., 1990. Growth and phosphorus acquisition of karri (*Eucalyptus diversicolor* F. Muell.) seedlings inoculated with ectomycorrhizal fungi in relation to phosphorus supply. *New Phytol.* **114**: 77–85.

Bowen, G. D., 1973. Mineral nutrient relations of ectomycorrhizae. Pp. 151–205 *in* Ectomycorrhizae — their ecology and physiology ed by G. C. Marks and T. T. Kozlowski. Academic Press: New York.

Brown, J. M. and Hopkins, A. J. M., 1983. The kwongan (sclerophyllous shrublands) of Tutanning Nature Reserve, Western Australia. *Aust. J. Ecol.* **8**: 67–71.

Burgman, M. A. and Lamont, B. B., 1992. A stochastic model for the viability of *Banksia cuneata* populations: environmental, demographic and genetic effects. *J. Appl. Ecol.* (in press).

Chesson, P. L. and Huntly, N., 1988. Community consequences of life-history traits in a variable environment. *Ann. Zool. Fennici* **25**: 5–16.

Chilvers, G. A. and Pryor, L. D., 1965. The structure of eucalypt mycorrhizas. *Aust. J. Bot.* **13**: 245–59.

Chilvers, G. A., 1973. Mycorrhizas and problems in association in *Eucalyptus* L'Herit. Ph.D. Thesis, Australian National University: Canberra.

Christensen, P. E. S., 1980. The biology of *Bettongia penicillata* (Gray, 1937) and *Macropus eugenii* (Desmarest, 1817) in relation to fire. *Forests Depart. West. Aust. Bull.* **91.**

Collins, B. G. and Robelo, T., 1987. Pollination biology in the Proteaceae of Australia and southern Africa. *Aust. J. Ecol.* **12**: 95–129.

Commonwealth of Australia, 1992. A national strategy for the conservation of Australia's biological diversity. Dept. Arts, Sport, the Environment and Territories: Canberra.

Corbeil, J. C. and Manser, M., 1988. Visual dictionary. Oxford: New York.

Cunningham, S. A., 1991. Experimental evidence for pollination of *Banksia* spp. by non-flying mammals. *Oecologia* **87**: 86–90.

Enright, N. J. and Lamont, B. B., 1989. Seed banks, fire season, safe sites and seedling recruitment in five co-occurring *Banksia* species. *J. Ecol.* **77**: 1111–22.

Enright, N. J. and Lamont, B. B., 1992. Survival, growth and water relations of *Banksia* seedlings in sand-mine tailings and adjacent scrub-heath sites. *J. Appl. Ecol.* (in press).

Erwin, T. L., 1991. An evolutionary basis for conservation strategies. *Science* **253**: 750–52.

Fineran, B. A. and Hocking, P. J., 1983. Features of parasitism, morphology and haustorial anatomy in loranthaceous root parasites. Pp. 205–27 *in* The biology of mistletoes ed by D. M. Calder and P. Bernhardt. Academic Press: Sydney.

Froend, R. H., 1987. Investigations into species richness patterns in the northern sandplain region of Western Australia. Ph.D. Thesis, University of Western Australia: Nedlands.

Gadgil, R. L. and Gadgil, P. W., 1971. Mycorrhiza and litter decomposition. *Nature* **233**: 133.

Gardner, J. H. and Malajczuk, N., 1988. Recolonization of mycorrhizal fungi of rehabilitated bauxite minesites in Western Australia. *For. Ecol. Manage.* **24**: 27–42.

George, A. S., 1981. The genus *Banksia* L.f. (Proteaceae). *Nuytsia* **3**: 239–473.

George, A. S., Hopkins, A. J. M. and Marchant, N. G., 1979. The heathlands of Western Australia. Pp. 211–30 *in* Ecosystems of the world. 9A. Heathlands and related shrublands ed by R. L. Specht. Elsevier: Amsterdam.

Grierson, P. F. and Attiwell, P. M., 1989. Chemical characteristics of the proteoid root mat of *Banksia integrifolia*. *Aust. J. Bot.* **37**: 137–43.

Hellmuth, E. O., 1971. Ecophysiological studies on plants in arid and semi-arid regions in Western Australia. *J. Ecol.* **59**: 365–74.

Herrara, R. T., Mérida, T., Stark, N. and Jordan, C. F., 1978. Direct phosphorus transfer from litter to roots. *Naturwissenschaften* **65**: 208–09.

Hobbs, R. J. and Atkins, L., 1991. Interactions between annuals and woody perennials in a Western Australian nature reserve. *J. Veg. Sci.* **2**: 643–54.

Hopkins, A. J. M. and Hnatiuk, R. J., 1981. An ecological survey of the kwongan south of Eneabba, Western Australia. *Wildl. Res. Bull. West. Aust.* **9.**

Hopper, S. D., 1979. Biogeographical aspects of speciation in the south-western Australian flora. *Ann. Rev. Ecol. Syst.* **10:** 399–422.

Houston, T. F., Lamont, B. B. and Radford, S., 1992. Co-evolution between a euryglossine bee and its food plant, *Verticordia nitens*. Pp. 329–34 *in* Proceedings of the sixth international conference on mediterranean climate ecosystems ed by C. A. Thanos. Athens: Greece.

Lamont, B., 1972. The effect of soil nutrients on the production of proteoid roots by *Hakea* species. *Aust. J. Bot.* **20:** 27–40.

Lamont, B., 1982. Mechanisms for enhancing nutrient uptake in plants, with particular reference to mediterranean South Africa and Western Australia. *Bot. Rev.* **48:** 597–689.

Lamont, B., 1983. Mineral nutrition of mistletoes. Pp. 185–204 *in* The biology of mistletoes ed by D. M. Calder and P. Bernhardt. Academic Press: Sydney.

Lamont, B. B., 1984. Specialized modes of nutrition. Pp. 126–45 *in* Kwongan: plant life of the sandplain ed by J. S. Pate and J. S. Beard. University of Western Australia Press: Perth.

Lamont, B., 1985a. Dispersal of the winged fruits of *Nuytsia floribunda* (Loranthaceae). *Aust. J. Ecol.* **10:** 187–93.

Lamont, B., 1985. The significance of flower colour change in eight co-occurring shrub species. *Bot. J. Linn. Soc.* **90:** 145–55.

Lamont, B., 1989. Biotic and abiotic interactions in *Banksia* woodland. *J. Roy. Soc. West. Aust.* **71:** 99–100.

Lamont, B. B., 1992a. Mineral nutrient relations in mediterranean regions of California, Chile and Australia. In press *in* Mediterranean-type ecosystems of the Pacific Basin: similarities and differences ed by M. T. Kalin de Arroya, P. H. Zedler and M. D. Fox. Springer-Verlag: New York.

Lamont, B. B., 1992b. Triangular trophic relationships in mediterranean Western Australia. Pp. 171–76 *in* Proceedings of the sixth international conference on mediterranean climate ecosystems ed by C. A. Thanos. Athens: Greece.

Lamont, B. B. and Bergl, S. M., 1991. Water relations, shoot and root architecture, and phenology of three co-occurring *Banksia* species: no evidence for niche differentiation in the pattern of water use. *Oikos* **60:** 291–98.

Lamont, B. B., Collins, B. G. and Cowling, R. M., 1985a. Reproductive biology of the Proteaceae in Australia and Southern Africa. *Proc. Ecol. Soc. Aust.* **14:** 213– 24.

Lamont, B. B., Connell, S. W. and Bergl, S., 1991a. Population and seed bank dynamics of *Banksia cuneata*: the role of time, fire and moisture. *Bot. Gaz.* **152:** 114–22.

Lamont, B. B., Downes, S. and Fox, J. E. D., 1977. Importance-value curves and diversity indices applied to a species-rich heathland in Western Australia. *Nature* **265:** 438–41.

Lamont, B. B., Enright, N. J. and Bergl, S. M., 1989. Coexistence and competitive exclusion of *Banksia hookeriana* in the presence of congeneric seedlings along a topographic gradient. *Oikos* **56:** 39–42.

Lamont, B. and Grey, J., 1984. Ants, extrafloral nectaries and elaiosomes on a pioneer species. Pp. 89–90 *in* Proceedings of the fourth international conference on mediterranean ecosystems ed by B. Dell. University of Western Australia: Nedlands.

Lamont, B. B., Hopkins, A. J. M. and Hnatiuk, R. J., 1984. The flora — composition, diversity and origins. Pp. 27–50 *in* Kwongan — plant life of the sandplain ed by J. S. Pate and J. S. Beard. University of Western Australia Press: Nedlands.

Lamont, B. and Kelly, W., 1989. The relationship between sclerophylly, nutrition and water use in two species from contrasting soils. Pp. 617–21 *in* Proceedings of the fifth international conference on mediterranean ecosystems ed by F. di Castri, C. Floret, S. Rombal and J. Roy. International Union of Biological Sciences: Paris.

Lamont, B. B., Le Maitre, D., Cowling, R. M. and Enright, N. J., 1991b. Canopy seed storage in woody plants. *Bot. Rev.* **57**: 277–317.

Lamont, B. B., Ralph, C. S. and Christensen, P. E. S., 1985b. Mycophagous marsupials as dispersal agents for ectomycorrhizal fungi on *Eucalyptus calophylla* and *Gastrolobium bilobum*. *New Phytol.* **101**: 651–56.

Lamont, B. B. and van Leeuwen, S. J., 1988. Seed production and mortality in a rare *Banksia* species. *J. Appl. Ecol.* **25**: 551–59.

Lamont, B. B. and Witkowski, E. T. F., 1992. Recruitment of four co-occurring *Banksia* species after spring and autumn fires at Hopetown. Final report to Dept. Conservation and Land Management and WA Bushfires Board. School of Biology, Curtin University: Western Australia.

Lamont, B. B., Witkowski, E. T. F. and Enright, N. J., 1992. Post-fire litter microsites: safe for seeds, unsafe for seedlings. *Ecology* (in press).

Low, A. B. and Lamont, B. B. 1990. Aerial and below-ground phytomass of *Banksia* scrub-heath at Eneabba, south-western Australia. *Aust. J. Bot.* **38**: 351–59.

MacFall, J., Slack, S. A. and Iyer, J., 1991. Effects of *Hebeloma arenosa* and phosphorus fertility on root acid phosphatase activity of red pine *(Pinus resinosa)* seedlings. *Can. J. Bot.* **69**: 380–83.

Macquarie Dictionary, 1981. Macquarie Library, St Leonards: New South Wales.

Malajcuk, N. and Hingston, F. J., 1981. Ectomycorrhizae associated with jarrah. *Aust. J. Bot.* **29**: 453–62.

Monk, D., Pate, J. S. and Loneragan, W. A., 1981. Biology of *Acacia pulchella* R.Br. with special reference to nitrogen fixation. *Aust. J. Bot.* **29**: 579–92.

Mosse, C. E., Powell, C. L. and Hayman, D. S., 1976. Plant growth responses to vesicular-arbuscular mycorrhiza. IX. Interactions between VA mycorrhiza, rock phosphate and symbiotic nitrogen fixation. *New Phytol.* **76**: 331–42.

Newland, C. E. and Wooller, R. D., 1985. Seasonal changes in a honeyeater assemblage in *Banksia* woodland near Perth, Western Australia. *N.Z. J. Zool.* **12**: 631–36.

Paine, R. T., 1969. A note on trophic complexity and community stability. *Amer. Nat.* **103**: 91–93.

Pearman, G. I. (ed), 1988. Greenhouse — planning for climatic change. CSIRO: Melbourne.

Rodriguez-Barrueco, C., MacKintosh, A. H. and Bond, G., 1970. Some effects of combined nitrogen on the nodule symbioses of *Casuarina* and *Ceanothus*. *Pl. Soil* **33**: 129–39.

Shea, S. R. and Kitt, R. J., 1976. The capacity of jarrah forest native legumes to fix nitrogen. *For. Dept. West. Aust. Res. Paper* **21**.

Stoutamire, W. P., 1975. Pseudocopulation in Australian terrestrial orchids. *Am. Orchid Soc. Bull.* **44**: 226–33.

Stoutamire, W. P., 1983. Wasp-pollinated species of *Caladenia* (Orchidaceae) in south-western Australia. *Aust. J. Bot.* **31**: 383–94.

Trappe, J. M., 1977. Selection of fungi for ectomycorrhizal inoculation in nurseries. *Ann. Rev. Phytopathol.* **15**: 203–22.

Wills, R. T., 1992. The ecological impact of *Phytophthora cinnamomi* in the Stirling Range National Park, Western Australia. *Aust. J. Ecol.* (in press).

Witkowski, E. T. F. and Lamont, B. B., 1991. Leaf specific mass confounds leaf density and thickness. *Oecologia* **88**: 486–93.

Witkowski, E. T. F., Lamont, B. B. and Connell, S. J., 1991. Seed bank dynamics of three co-occurring banksias in south coastal Western Australia: the role of plant age, cockatoos, senescence and interfire establishment. *Aust. J. Bot.* **39**: 385–97.

Witkowski, E. T. F., Lamont, B. B., Walton, C. S. and Radford, S., 1992. Leaf demography, sclerophylly and ecophysiology of two *Banksia* species with contrasting leaf life spans. *Aust. J. Bot.* (in press).

Zohar, Y., Waisel, Y. and Korschan, R., 1981. Heat and cold resistance of *Eucalyptus occidentalis* Endl. leaves and its relationship to soil water conditions. *Aust. J. Ecol.* **6**: 79–84.

PLATE 4

ABOVE: *Ctenophorous maculatus,* a common compotent of the lizard fauna on Durokoppin Nature Reserve, north of Kellerberrin, Western Australia. Reptiles seem to have been relatively little affected by fragmentation. (Photo: R. J. Hobbs).

BELOW: The lizard *Moloch horridus,* on Durokoppin Nature Reserve, north of Kellerberrin, Western Australia. (Photo: D. Steven).

CHAPTER 7

The role of faunal diversity in ecosystem function

Robert J. Lambeck[1]

Faunal biodiversity contributes to the maintenance of primary ecosystem processes by providing an array of pathways for the flux of nutrients, water and energy. Changes to biodiversity alter the number of pathways through which nutrients and energy move, resulting in altered rates of cycling of essential elements.

The impact of changes in faunal biodiversity will depend upon whether functionally equivalent species are present, and whether they are able to compensate for changes in the distribution and abundance of analogous species. Compensatory responses may occur where the distribution and abundance of functional analogues are determined by synecological relationships such as competition and predation, but will not occur where autecological factors determine distributions.

Biodiversity contributes to the maintenance of ecosystem processes in the face of environmental fluctuations, and increases the time and area over which these processes occur. Faunal biodiversity can increase the functional stability of a region and enhance the probability of maintaining or restoring function, following local extinctions. In the fragmented landscapes which typify much of Australia's mediterranean regions, this capacity has been significantly reduced.

Introduction

Biodiversity encompasses the heterogeneity which occurs at all levels in a region's biota (Hobbs, Chapter 1). It includes not only the variety of species of plants, animals and other organisms, but also the genes they contain and the communities and ecosystems

[1]School of Environmental Biology, Curtin University of Technology, PO Box U1987, Perth, Western Australia, Australia 6001.

Pages 129–48 *in* BIODIVERSITY OF MEDITERRANEAN ECOSYSTEMS IN AUSTRALIA ed by Richard J. Hobbs. Surrey Beatty & Sons, 1992.

that they make up. Patterns of biodiversity form a nested hierarchy, with genotypic variability forming the basis upon which diversity at the population, species, assemblage, and ecosystem levels is built.

Similarly, ecosystem functioning may be considered to be hierarchical (Hobbs, Chapters 1, 11), comprising the primary processes of nutrient, energy and water cycling, as well as various "sub-functions" in which fauna may play a special role, such as pollination and propagule dispersal.

The relationship between faunal biodiversity and ecosystem function is expressed at all levels of the biodiversity hierarchy. Individual differences within species will be reflected in different capacities to survive environmental stresses or other disturbances. Differences between species affect the rates of nutrient and energy flow within a community, as well as the pathways via which these elements are mobilized. For example, different herbivore species will forage on different plant types, and different predators will consume alternative prey items. The interactions between the various species, together with their responses to local environmental conditions, will result in differences in species assemblages at different locations, and subsequent differences in regional patterns of nutrient and energy flux.

The relationship between biodiversity and function becomes increasingly more complex at higher levels of the biodiversity hierarchy. It is much easier to identify the functional significance of the individual components of biodiversity (for example the role of a particular species), than it is to identify the significance of biodiversity *per se*. The functional significance of biodiversity is likely to be greater than the sum of the contributions of the individual components which make up that diversity. For example, the removal or addition of any individual organism will result in some small change in the rate of a given process. However, the cumulative impact of a succession of such changes is not simply the sum of the impacts associated with the removal or addition of each individual. The impact of losses or additions at any level of the biodiversity hierarchy will also be a consequence of how other species respond to such changes in abundance.

Analysis of the role of biodiversity in ecosystem functioning must therefore consider more than the role of the individual components of that diversity. It must take into account the interactions between the various taxa, and how the different components of the region's biodiversity respond to changes in the abundance of other members of the community.

Many of the general principles which underlie the relationship between faunal biodiversity and ecosystem function in mediterranean ecosystems will also apply to other ecosystems, and will be relevant to plant communities. In order to address these issues I will use a framework of faunal assemblages which are structured by trophic relationships between the constituent organisms, by flows of energy and nutrients through these food webs, and by the grouping of the component species into guilds which perform functionally equivalent roles.

This chapter explores the relationships between faunal biodiversity and ecosystem function by focusing primarily on the species and population levels of biodiversity, as these are the scales at which the majority of faunal studies in the region have been conducted. Studies in the Australian mediterranean regions which have directly addressed such questions, or which have indirectly shed light upon these issues, are few. Consequently this chapter does not constitute a review of the role of faunal biodiversity in ecosystem functioning in this region. Instead, it examines the relationships between faunal biodiversity and ecosystem function from a theoretical perspective with examples drawn from Australia's mediterranean-climate regions where these are available.

Faunal biodiversity in the Australian mediterranean-climate regions

The mediterranean-climate regions of Australia (as described in Chapter 1 of this volume) are amongst the most modified regions of the continent. They have been heavily exploited for pastoral, agricultural and timber production for over 100 years (see Saunders and Hobbs, Chapter 4). The resultant loss or alteration of habitat, together with the introduction of exotic predators and competitors, has resulted in significant changes to the fauna of the region. In the wheatbelt of Western Australia alone, 13 species of mammals have become locally extinct (Kitchener *et al.* 1980), and in the central wheatbelt 15 of 131 previously recorded bird species no longer occur, with an additional 24 bird species undergoing a reduction in abundance and/or range (Saunders 1989; Saunders and Curry 1990; Saunders and de Rebeira 1991; see also Saunders and Hobbs, Chapter 4). While there is no evidence that these land-use practices have caused extinctions of invertebrate species, it is clear that they do result in substantial changes to local invertebrate diversity (Scougall *et al.* 1993; Springett 1976). Consequently the current patterns of faunal biodiversity no longer reflect those which occurred before the arrival of Europeans. Prior to this, Aboriginal use of fire may also have altered patterns of floral and faunal diversity (Horton 1982).

The vertebrate fauna of the Australian mediterranean-climate region is characterized by a relatively low level of endemism (Table 1). Many of the species which have been recorded as occurring in the region are representatives of either arid zone or more mesic faunas, and the mediterranean-climate region represents the limits of their distributions. Such species are commonly confined to the periphery of the region. This pattern reflects the transitional nature of this climatic zone between the dryer inland areas and the more mesic environments to the east and south-east.

Table 1. The extent of endemism in Australia's mediterranean-climate zone vertebrate species (data from Cogger 1986; Schodde 1981; Strahan 1983).

	Number of species in Australian mediterranean	Percentage of Australian fauna	Number endemic	Percentage of mediterranean species which are endemic
Mammals	82	32	16	19.5
Birds	163	36	9	6
Snakes	43	33	11	25
Lizards	144	29	42	29
Frogs	36	20	23	64

The low level of endemism suggests that the distribution of many vertebrate species in the region may largely be a consequence of historical and biogeographic factors, rather than a result of adaptations to local ecological conditions. Much of the vertebrate fauna has broad environmental tolerances which enables it to persist under a wide range of conditions, including those presented by the marked seasonal climate of mediterranean regions. Consequently, the functional significance of vertebrate faunal biodiversity in the region is less likely to lie in specialized relationships between the various components of the biota. Rather, much of the fauna could be viewed as having a generalized role which would also apply in the other regions in which it occurs.

The low level of endemism is particularly conspicuous for birds. Only nine species are endemic to the mediterranean-climate zone (Schodde 1981), although over 190 species have been recorded in the Western Australian region alone (Saunders 1992). Of over 80 species of mammals recorded from the region only 16 or approximately 20 per cent are confined to it, while approximately 28 per cent of the mediterranean zone reptiles are endemic. Frogs provide a conspicuous exception to this trend with 64 per cent of the local species being restricted to the region (data from distribution maps of Strahan 1983; Cogger 1986). Many invertebrate groups also show a higher degree of endemism which to a large extent reflects the dispersal characteristics of particular taxonomic groups. More vagile groups such as butterflies and grasshoppers have a higher proportion of species which extend beyond the region, whereas many terrestrial and epigaeic arthropods such as scorpions and mites have very restricted distributions (Mark Harvey, pers. comm.).

Faunal biodiversity and function at the species level

Food webs and functional analogues

The functional significance of a region's flora could be considered to lie in its mediating role between the biotic and abiotic elements of the environment. Plants transform non-living elements into the living tissue that forms the basis of biodiversity. The importance of the fauna however is based primarily on interactions between the biotic components of an ecosystem. A major functional consequence of faunal activity is the mobilization of the nutrients and energy which are "locked up" in other organisms. Animals serve to modify the rate at which elements circulate within a system and the pathways via which they move.

If we adopt a view of ecosystem function based upon complex webs of alternative nutrient, water and energy pathways through a community, each mediated by a different suite of organisms, it becomes apparent that the loss of one or more of these pathways (due to the loss of their constituent organisms) may not necessarily result in a cessation of function. Rather, such a loss will influence the rate and manner in which elements cycle, and hence alter the contribution of that community to overall ecosystem function. Such a view immediately identifies one of the primary consequences of biodiversity: different levels of diversity provide an array of species which mediate the flux of nutrients and energy. Changes in biodiversity can alter the number of pathways, and consequently the rate of flow of nutrients and energy through the food webs which make up the community. For example, Springett (1976) found that the reduced number of species making up the soil microfauna of pine plantations, relative to that of adjoining native vegetation, was associated with a reduction in the rate of litter decomposition.

The impact of changes in biodiversity will, in part, be a consequence of the number of alternative species in the system which are carrying out equivalent roles. Such species are termed functional equivalents or functional analogues (Barbault *et al.* 1991). The importance of these analogous species depends upon the manner in which they respond to the demise or addition of other functionally equivalent species. In a manner equivalent to Gause's (1934) contention that two ecologically equivalent species cannot live together in the same place, so species which are essentially carrying out the same function will tend to be temporally or spatially separated as a consequence of either autecological factors, such as different habitat or resource requirements, or synecological processes such as predation, parasitism and competition. This distinction between autecological and synecological determinants of distribution and abundance is important because it significantly affects the consequences of the loss of a given species.

Synecological Processes

If the distribution and abundance of two functionally analogous species is determined by processes such as competitive interactions between them, then the loss of one (for example, species E_2 in Fig. 1) may be compensated for by competitive release (Cody 1986), resulting in an increase in the abundance of the other species (species E_3, Fig. 1). The result of such density compensation may be no net loss of function. If the increased contribution of the persisting species completely compensates for the removal of the other species, then they could be considered to be complete functional equivalents. A species whose loss will be fully compensated for by the expansion of another could be considered to be functionally redundant. The likelihood of such total compensation occurring in natural ecosystems is extremely remote. Species which may be functionally equivalent in one aspect of their biology will not necessarily be so in others. For example, two species which may be equivalent in terms of resources they consume may serve as food resources for a different array of predators. Even if two species are functionally equivalent under prevailing conditions, this equality may not persist under altered circumstances, such as human induced habitat or climate changes.

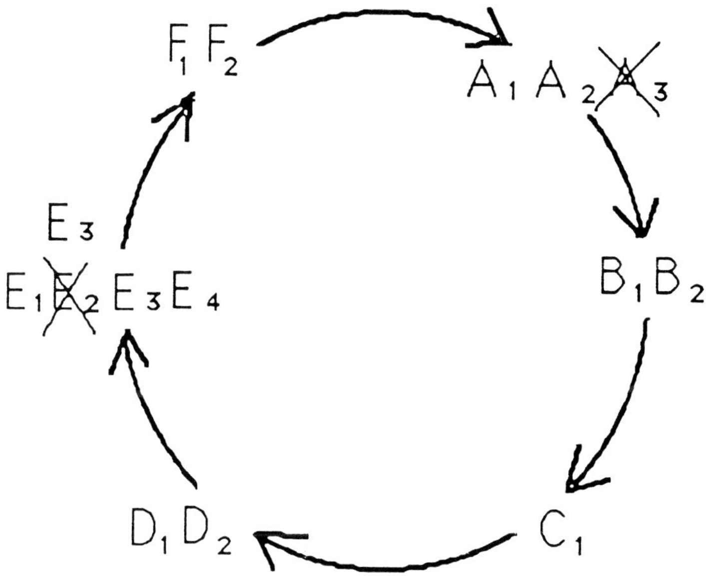

Fig. 1. The role of diversity in the cycling of nutrients and energy. Letters represent functional groups with subscripts denoting functionally analogous species. Functional compensation occurs where one species (E_3) increases in abundance in response to the decline of another (E_2). Where analogous species have different habitat requirements, such compensation will not occur when a species (e.g., A_3) is lost, resulting in a reduced rate of function. The loss of a species with no functional analogues (C_1) will seriously impede ecosystem processes in which it participates.

In order to determine the importance of species diversity in the maintenance of ecosystem functioning, it is necessary to ascertain whether species which compensate for changes in the abundance of each other are in fact functional analogues. That is, does density compensation necessarily equate with functional compensation. Fox and Pople (1984) demonstrated that density compensation occurs between two species of small mammals in heathland in Eastern Australia. Experimental removals and additions of *Pseudomys novaehollandiae* resulted in reciprocal fluctuations in the abundance of the introduced house mouse, *Mus musculus*. The functional significance of these species in the community was not assessed, so it is not clear that the observed density compensation resulted in functional compensation. However, a high degree of dietary overlap suggests that fluctuations in the rates of resource consumption may be reduced by the reciprocal density adjustments of the two species. A decline in the numbers of one species in the absence of the other could result in altered rates of resource consumption which, depending on the nature of the resource, could have further implications. For example, a changed rate of seed consumption due to the loss of a granivorous species may alter recruitment rates for some plant species.

In another example, Springett (1976) found that a reduction in the number of species of litter and soil invertebrates in pine plantations, relative to adjoining native vegetation, did not result in a decline in the total number of individuals. This apparent density compensation did not however, equate with functional compensation. Although the number of individuals remained constant, the overall rate of decomposition was lower where there were fewer species.

Conversely, the maintenance of a given function following a decline in the number of species is not necessarily indicative of functional compensation. In a comparative study of two sites in Jarrah *(Eucalyptus marginata)* woodland in Western Australia, one of which had been affected by Jarrah dieback *(Phytophthora cinnamomi)* and one unaffected, Postle *et al.* (1986) found similar rates of litter decomposition in both sites in spite of significantly fewer taxa of decomposers, and fewer individuals, in the dieback affected area. The apparent functional compensation in the dieback affected area was attributed to increased physical breakdown of litter in the modified site due to thinning of the canopy which left the litter more exposed to abiotic weathering processes (see also Majer, Chapter 9).

Such examples illustrate that the increase in abundance of one species in response to the decline of another, or the maintenance of function in the face of changed diversity, does not necessarily mean that functional compensation has occurred. Such changes in abundance may simply be a consequence of species occupying some other dimension of niche space which has become vacant as a result of the reduced diversity.

Density compensation need not be confined to species expansions in the face of reduced biodiversity. Compensatory processes could also include species contractions in response to invasions. An introduced or invading species may not cause a change in function if an increase in its numbers causes a corresponding decline in numbers of a functional analogue. For example the introduction of rabbits to Australia (see Hobbs, Chapter 1) coincided with a decline in marsupial herbivores, and may have over-compensated for any loss of function attributable to the reduced diversity of native species. In this case, a reduction in species richness has resulted in an increase in the rate of a given function (i.e., herbivory) and accelerated the conversion of nutrients and energy from plant to animal biomass.

Autecological processes

If the distribution and abundance of a member of a functional group is not limited by the presence of analogous species, but is determined by different habitat or resource requirements, then a decline in the numbers of that species (e.g., species A_3, Fig. 1) will not be compensated for by a corresponding increase in the numbers or activity of the alternative species. In such a case one would expect to see a net decline in rates of nutrient and energy cycling. In the absence of analogues which can compensate for the decline, these functional changes will occur regardless of the level of biodiversity in the community. Higher levels of diversity simply increase the probability of compensatory analogues being present.

Although autecologically distributed species do not compensate for altered function due to changes in biodiversity, a range of such species can have important consequences. By functioning at different temperatures, different moisture regimes, or different times of the day or year, an array of analogous species, in a range of different habitats, will contribute to enhanced productivity. By increasing the likelihood of asynchronous function, biodiversity has the capacity to dampen fluctuations in a given ecosystem process (Fig. 2). Studies by Majer and Koch (1982) illustrate the contribution of faunal diversity to seasonal continuity of function in epigaeic invertebrate communities (Table 2). Although the majority of herbivorous species are active in the austral summer, some species remain active at other times of the year. While the number of species active (and hence the rate of decomposition) may decline significantly over the winter months, the diversity of decomposers ensures that there is at least some invertebrate activity throughout the year. Presumably it is this biodiversity at lower trophic levels which supports the observed year-round presence of predatory species.

Similarly, termite diversity can generate both spatial and temporal continuity of decomposition. Seasonal differences in the activity of different termite species results in the presence of active termites throughout the year in remnants of native vegetation in the Western Australian wheatbelt (Abensperg-Traun 1991). A

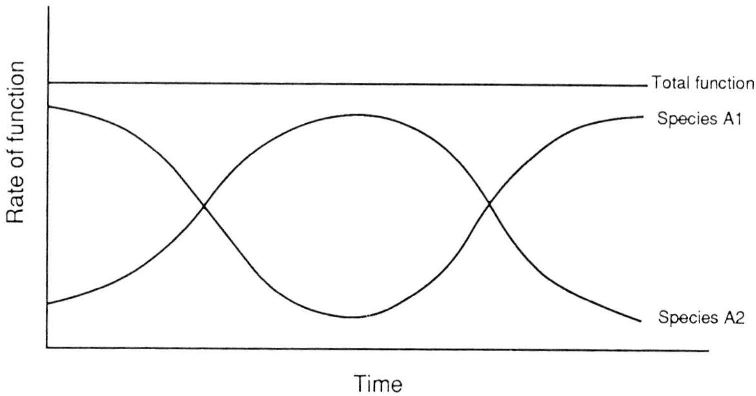

Fig. 2. The contribution of diversity to the stability of ecosystem processes. Functional analogues which are active asynchronously will result in a given function continuing at a relatively constant rate, even though the numbers of the participating species may fluctuate.

variety of termite species also results in foraging activity in a range of different habitat types, with heath species harvesting grasses, and woodland species consuming fallen wood (Abensperg-Traun and de Boer 1990).

The significance of an array of termite species is further enhanced by their additional role as prey items for a wide range of predators (Abensperg-Traun, pers. comm.). Spatial and temporal differences in activity between species enhances their reliability as a food resource and supports a higher diversity of predators. The high species richness of scincid lizards of the genus *Ctenotus* in Australia's arid zone has been attributed to the variety of termite species present, with different lizard species consuming different termite genera or foraging guilds (James 1991). Since both termites and termite predators constitute a conspicuous component of the mediterranean-climate fauna, their role in this region could be expected to be similar to that in the arid zone.

Keystone Species

The importance of a given species in overall ecosystem function is not simply due to the particular attributes of that species, but rather its importance is dependent upon what other functionally analogous species are present in the community, and how these species respond to its demise. Keystone species (Paine 1974; Bond 1993) can be viewed as species which make a major contribution

Table 2. The contribution of invertebrate diversity to the time over which ecosystem processes occur. Numbers indicate the number of individuals caught in pitfall traps. (Manjimup data from Majer and Koch 1982.)

Taxa	O	N	D	J	F	M	A	M	J	J	A	S
PREDATORS												
Dermaptera	16	17	21	84	27	12	8	11	5	2	3	3
Promecoderus	2	8	19	9	3	11	3	5	6			2
Simodontus	2			3	1	4	2	11	6	5	2	5
Quedius			2					26	2	2		2
Anotylus	13	3						1				
Aleocharinae	3	5										
Reduviidae								1				
Nerthra					4							
Carabidae larvae	2							1				
Pselaphidae								4				3
Pompilidae					1							
HERBIVORES												
Phaulacridium					2							
Ledrinae					1							
Semanopterus					1							
Byrrhomorpha					1							
Conoderus					2							
Cherrus					1							
Biphyllocera				1	1							
Hylaeus				1	1							
Neophylloctocus				8	2	1						
Cydnus			1	1	4							
Acantholopus			1		1							
Leptocorisa					4						1	
Calolampra					2	3	7	12	19	1		3
Gryllidae				2			2	3	2			
Lepidoptera			1									
Apion		1										
Laxta		1										
Curculionidae						1						
Tranes						2						
Aedrioides									1			
Dialeptops											1	
DECOMPOSERS												
Antichiropus (b)		1										
Armidillididae		2	14	5		3	2					
Laevophiloscia (c)		4		2	21	2						
Laevophiloscia (a)		2				3	2		1		2	

to the cycling of nutrients and energy but which do not have functional analogues (e.g., species C_1 in Fig. 1). Such species represent a functional bottleneck because they mediate a disproportionate amount of nutrient and energy flux, and consequently may limit the rates of particular processes and hence the productivity of a system. The loss of such species can result in major changes to the function, and hence composition, of the community in which they occur.

Because of the unspecialized nature of much of the fauna, animal examples of keystone species from the Australian mediterranean region are uncommon. Even where there are interdependencies (for example, monolectic, or species specific relationships between orchids and wasps (Stoutamire 1983)), the number of species involved are few and the repercussions of their loss are limited.

Plant species are more likely to play keystone roles in this region (Lamont, Chapter 6). For example, in the central wheatbelt of Western Australia, *Banksia prionotes* provides an important nectar source for honeyeaters for a limited period of the year when there are virtually no other nectar-producing species in flower (Lambeck, unpubl. data). Although this species is widespread and relatively abundant over much of its range, it is poorly represented in the central wheatbelt and extremely localized. Because of its asynchronous flowering relative to other nectar producing species, it provides a critical resource for honeyeaters which, as pollinators, play an important role in the maintenance of floral biodiversity in the region. Many of the region's plants use honeyeaters as vectors for pollen transfer (Keighery 1982) and a reduction in bird numbers could have a significant impact on plant reproductive success. For example, *B. prionotes* will not set seed unless pollinated by birds (Collins and Spice 1986), and although experimental exclusion of birds from *Hakea scoparia* did not prevent seed set, it resulted in a reduction in the number of follicles formed (Lambeck, unpubl. data). The loss of such nectar producing plants could have a significant impact on the capacity of honeyeaters to persist in the region. *B. prionotes* has become a keystone mutualist (Gilbert 1980) not solely because of its particular attributes, but because of its relationship to other species in the region. The absence of functional analogues has elevated its importance. This species does not play the same critical role in coastal parts of its range, where it occurs in sympatry with the synchronously flowering *Banksia menziesii*.

Faunal biodiversity and function at the population level

Biodiversity not only encapsulates the number of different categories (species, populations, etc.) in a region, but also includes the number and distribution of individuals within those groups. Biodiversity is commonly equated with "species richness" (e.g., Koshland 1991). However, the impact of biodiversity on function is dependent not only upon the number of species present, but also upon the number of individuals of those species, and their distribution in the landscape.

As was the case for changes in the number of species, the impact of a change in the abundance of a species will also be a consequence of whether functional analogues are present and how they respond to the change. Changes in ecosystem processes are possible without changes in species richness, simply through fluctuations in the abundance of the constituent species. Any changes in function that may have occurred in Australian mediterranean landscapes over the past 150 years are more likely to be due to reductions in the area of natural habitat available, and consequent reductions in population sizes of the constituent species, rather than to a loss of species. Land clearing in Australia's mediterranean-climate regions has removed from 60 per cent to over 90 per cent of the natural bush (Beard and Sprenger 1984; Saunders and Hobbs, Chapter 4; Majer, Chapter 9). This clearing has had a significantly greater impact upon the numbers of individuals of the species present, and their patterns of distribution, than on the numbers of species. For example, the mammal fauna in the wheatbelt of Western Australia is represented by some 43 species (Kitchener *et al.* 1980) suggesting a relatively rich fauna, but of these only 12 species are moderately common to abundant. The contribution made by the majority of native mammal species to the mobilization of nutrients and energy will therefore be limited and probably localized. It is the decline in the biomass of flora and fauna, rather than the reduced number of species, that has contributed most to changes in the rates and area over which ecosystem processes proceed. A community of high species richness, but with low population densities, may well be less functional than one with fewer species but more individuals. Loss of diversity at the population level may therefore present a greater threat to ecosystem function than does the loss of diversity at the species level. The demise of a given species is simply the failure of the last population of that species, and any change of function would have commenced with the first decline in numbers. The significance of species loss therefore is the removal of future options for recovery, whereas the greatest threat to function is the loss of local populations.

Faunal diversity, complexity and stability

While community diversity is a measure of the variety, number and distribution of the animals present, complexity reflects the degree of connectance between the various species. Connectance may take the form of vertical linkages between different trophic levels, or horizontal linkages between species on the same trophic level (Pimm 1984). The vertical linkages reflect pathways by which nutrients and energy flow through a community, and are typically mediated by producer/consumer, or predator/prey relationships. Horizontal linkages on the other hand represent non-trophic interactions such as competition between functional analogues, and the specialized interactions involving symbionts and mutualists such as seed dispersers and pollinators.

The role of diversity and complexity in the maintenance of community stability has been the subject of considerable debate, with conflicting views being expressed. MacArthur (1957) argued that the more pathways available for energy to reach a consumer, the less severe would be the consequences of the failure of a given pathway. Elton (1958) also contended that increasing complexity should generate greater stability. By contrast, May (1984) argued, on the basis of theoretical models, that multi-species communities turn out to be less stable as more species are added. Cody (1986) also found that high diversity in mediterranean-climate regions tends to be associated with a greater probability of rarity, and that species in more diverse communities have a greater risk of extinction.

It is important, in this context, to examine the relationship between species extinction probabilities and threats to functional stability. For a given unit of area, a more diverse community will, on average, have fewer individuals per species. These smaller populations will consequently have a greater probability of local extinction (Pimm *et al.* 1988). In an unfragmented landscape such local extinction may not necessarily jeopardize ecosystem functioning. Such a landscape would be a dynamic one in which local extinctions would, in the longer term, be balanced by recolonization. While there may be intermittent changes in some functions following local extinctions, these would be negated over time as the relevant species recolonized the area. Additionally, a more diverse community will have a greater probability of containing functional analogues which may compensate for local changes in function. In other words, such diverse systems may be locally unstable in terms of species extinction probabilities, but may be able to retain functional stability because of a higher probability of short-term functional compensation, and long-term re-establishment of function through recolonization. A high probability of local extinction need not necessarily equate with functional instability.

In the highly fragmented landscapes which characterize much of the Australian mediterranean regions, options for functional compensation and recolonization, and hence much of the functional resilience (Majer, Chapter 9), have been lost. The patchy distribution of the original vegetation (Beard and Sprenger 1984; Specht 1972) is associated with a relatively high beta diversity. Animal species tend not to be uniformly distributed throughout these spatially variable environments. Fragmentation of such a landscape results in unequal representation of the various components of the biota in the fragments which remain. The probability of extinction is increased, due to the reduction in the number of species which can be supported by the reduced area (MacArthur and Wilson 1963; Connor and McCoy 1979), and because of pressures imposed by adjacent land-use practices. In addition, the stochastic risks associated with natural disturbance remain, increasing the probability of populations being reduced to levels where demographic fluctuations can jeopardize genetic integrity (Gilpin and Soulé 1986).

In these fragmented landscapes the combination of natural disturbances and human impacts has a much greater capacity to affect all of the members of a guild or functional group than is the case for continuous, unmodified environments. Any changes in function resulting from a change in species diversity will be less likely to be compensated for in the short term, as there will be fewer analogous species. Furthermore, the probability of long-term restoration of function through recolonization will be reduced as a result of the increased distance to source populations and the hostility of the intervening habitat.

The presence of a diverse fauna in unfragmented landscapes contributes to continuity of function following local extinction, thereby maintaining conditions suitable for recolonization. The change in function associated with fragmentation may eliminate the possibility of re-establishment of a population, even if dispersal back to these disturbed remnants is possible. Such remnants may become "sinks" (Howe *et al.* 1991) which dispersing organisms can colonize but in which they are either unable to survive or unable to successfully produce or raise young. Extinction will then become the dominant population process, resulting in further changes to the manner in which the ecosystem functions. In fragmented environments the probability of local extinction is higher, and such extinctions are more likely to result in permanent local changes to functioning than is the case for undisturbed habitat.

Conclusion

If ecosystem functioning is to be construed as the persistence of those processes necessary for the cycling of nutrients and energy, then changes in the level of faunal biodiversity will not result in the cessation of function. By altering the number of alternative pathways via which nutrients and energy can move through a given system, changes in faunal biodiversity will result in altered rates of function. Complete loss of function will occur only where catastrophic events or human land-use practices destroy the primary production and decomposition trophic levels (Ehrlich and Mooney 1983). The primary processes of nutrient, energy and water flux could theoretically persist in the absence of fauna. Elements extracted from the soil and air by plants could return as a result of senescence, rather than consumption by herbivores, and decomposition could result from the actions of microflora or physical weathering.

Suggesting that faunal biodiversity is not essential for ecosystem functioning is not equivalent to saying that animals are not important in the Australian mediterranean-climate regions. As mobilizers of nutrients and energy, and as vectors for pollen and seed dispersal, they play an important part in maintaining the "characteristic" patterns and rates of processes, and therefore influence the assemblages of species in the region. However, it is the particular attributes of the animals present, their relative abundances, and the nature of their interactions, rather than the number of species, which is most important in maintaining these rates of function. Maintenance of a given level of function is dependant upon the presence of appropriate species. The maintenance of biodiversity simply increases the probability of critical components being represented. High levels of biodiversity will not sustain a given level of function if critical components of that diversity are missing.

The impact of changes in biodiversity will be a consequence of the functional role of those components which have been added or removed, and the responses of other species to these changes.

If functional analogues which compensate for such changes are present, then the impact of the change may be reduced. Higher biodiversity increases the probability of their presence. Because functionally analogous species respond to different environmental parameters, biodiversity can increase the area and time over which ecosystem processes occur, and can provide a broader range of options for the persistence of process in the face of change.

If it is possible to have continued ecosystem function with reduced faunal biodiversity, albeit at different rates, why then need we be concerned about loss of diversity? Is an ecosystem with fewer species less functional than one with more species, or alternatively, is one in which nutrients and energy cycle more slowly, of lesser quality than one in which they cycle more rapidly? Different levels of biodiversity do not result in better or worse ecosystems, but simply in different ones. Biodiversity is a value-free concept until we, as the exploiters and modifiers of that biodiversity, impose value upon it. The importance of biodiversity in what is now a predominantly anthropogenic environment can only be defined relative to our expectations of that environment. These expectations need to be identified and prioritized before we can determine the patterns of biodiversity that are necessary to meet those expectations.

If, as suggested in this chapter, the ecological consequences of changes in faunal biodiversity are simply different levels of function, and that these are neither better nor worse until defined as such by us, then ecological significance also becomes dependant on human values. The "ecological" importance of biodiversity cannot be separated from the "other" values of commodity, amenity and morality (Norton 1988; Walker 1992). The ecological significance of biodiversity can only be defined relative to a given level of function, and in an anthropogenic landscape, this level will be determined by commodity, amenity and moral values. Once objectives have been defined by these criteria, an understanding of the ecological relationship between biodiversity and function will assist with the management of the biota in a manner which enables these objectives to be attained.

References

Abensperg-Traun, M., 1991. Seasonal changes in activity of subterranean termite species (Isoptera) in Western Australian wheatbelt habitats. *Aust. J. Ecol.* **16**: 331–36.

Abensperg-Traun, M. and de Boer, E. S., 1990. Species abundance and habitat differences in biomass of subterranean termites (Isoptera) in the wheatbelt of Western Australia. *Aust. J. Ecol.* **15**: 219–26.

Barbault, R., Colwell, R. K., Dias, B., Hawksworth, D. L., Huston, M., Laserre, P., Stone, D. and Younès, T., 1991. Conceptual framework and research issues for species diversity at the community level. Pp. 37–71 *in* From genes to ecosystems: a research agenda for biodiversity ed by O. T. Solbrig. IUBS: Paris.

Beard, J. S. and Sprenger, B. S., 1984. Geographical data from the vegetation survey of Western Australia. Vegetation Survey of Western Australia. Occasional Paper No. 2. Vegmap Publications: Perth.

Bond, W. J., 1993. Keystone species. *In* Ecosystem function of biodiversity ed by E.-D. Schulze and H. A. Mooney. Springer-Verlag: Heidelberg, in press.

Cody, M. L., 1986. Diversity, rarity, and conservation in mediterranean-climate regions. Pp. 122–52 *in* Conservation biology: the science of scarcity and diversity ed by M. E. Soulé. Sinauer Associates: Sunderland, Mass.

Cogger, H. G., 1986. Reptiles and amphibians of Australia. Reed: Sydney.

Collins, B. G. and Spice, J., 1986. Honeyeaters and the pollination biology of *Banksia prionotes* (Proteaceae). *Aust. J. Bot.* **34:** 175–85.

Connor, E. F. and McCoy, E. D., 1979. The statistics and biology of the species-area relationship. *Amer. Nat.* **113:** 791–833.

Ehrlich, P. R. and Mooney, H. A., 1983. Extinction, substitution, and ecosystem services. *BioScience* **33:** 248–54.

Elton, C. S., 1958. The ecology of invasions by animals and plants. Methuen: London.

Fox, B. J. and Pople, A. R., 1984. Experimental confirmation of interspecific competition between native and introduced mice. *Aust. J. Ecol.* **9:** 323–34.

Gause, G. F., 1934 (Reprinted 1964). The struggle for existence. Macmillan (Hafner Press): New York.

Gilbert, L. E., 1980. Food web organization and the conservation of neotropical diversity. *In* Conservation biology: an evolutionary-ecological perspective ed by M. E. Soulé and B. A. Wilcox. Sinauer Associates: Sunderland, Mass.

Gilpin, M. E. and Soulé, M. E., 1986. Minimum viable populations: processes of species extinction. Pp. 19–34 *in* Conservation biology: the science of scarcity and diversity ed by M. E. Soulé. Sinauer Associates: Sunderland, Mass.

Horton, D. R., 1982. The burning question: Aborigines, fire and Australian ecosystems. *Mankind* **13:** 237–51.

Howe, R. W., Davis, G. J. and Mosca, V., 1991. The demographic significance of "sink" populations. *Biol. Conserv.* **57:** 239–56.

James, C. D., 1991. Temporal variation in diets and trophic partitioning by coexisting lizards of the genus *Ctenotus* (Scincidae) in central Australia. *Oecologia* **85:** 553–61.

Keighery, G. J., 1982. Bird-pollinated plants in Western Australia. Pp. 77–89 *in* Pollination and evolution ed by J. A. Armstrong, J. M. Powell and A. J. Richards. Royal Botanic Gardens: Sydney.

Kitchener, D. J., Chapman, A., Muir, B. G. and Palmer, M., 1980. The conservation value for mammals of reserves in the Western Australian wheatbelt. *Biol. Conserv.* **18:** 179–207.

Koshland, D. E., 1991. Preserving biodiversity. *Science* **253:** 717.

MacArthur, R. H., 1957. Fluctuations of animal populations as a measure of community stability. *Ecology* **36:** 533–36.

MacArthur, R. H. and Wilson, E. O., 1963. An equilibrium theory of insular zoogeography. *Evolution* **17**: 373–87.

Majer, J. D. and Koch, L. E., 1982. Seasonal activity of hexapods in woodland and forest leaf litter in the south-west of Western Australia. *J. Roy. Soc. West. Aust.* **65**: 37–45.

May, R. M., 1984. An overview: real and apparent patterns in community structure. Pp. 3–18 *in* Ecological communities: conceptual issues and the evidence ed by D. R. Strong, D. S. Simberloff, L. G. Abele and A. B. Thistle. Princeton University Press: Princeton, New Jersey.

Norton, B., 1988. Commodity, amenity, and morality. The limits of quantification in valuing biodiversity. Pp. 200–05 *in* Biodiversity ed by E. O. Wilson. National Academy Press: Washington, DC.

Paine, R. T., 1974. Intertidal community structure: experimental studies on the relationship between a dominant competitor and its principle predator. *Oecologia* **15**: 93–120.

Pimm, S. L., 1984. Food chains and return times. Pp. 397–412 *in* Ecological communities: conceptual issues and the evidence ed by D. R. Strong, D. S. Simberloff, L. G. Abele and A. B. Thistle. Princeton University Press: Princeton, New Jersey.

Pimm, S. L., Jones, H. L. and Diamond, J. M., 1988. On the risk of extinction. *Amer. Nat.* **132**: 757–85.

Postle, A. C., Majer, J. D. and Bell, D. T., 1986. Soil and litter invertebrates and litter decomposition in Jarrah *(Eucalyptus marginata)* forest affected by Jarrah dieback fungus *(Phytophthora cinnamomi)*. *Pedobiologia* **29**: 47–69.

Saunders, D. A., 1989. Changes in the avifauna of a region, district and remnant as a result of fragmentation of native vegetation: the wheatbelt of Western Australia. A case study. *Biol. Conserv.* **50**: 99–135.

Saunders, D. A., 1992. The effect of habitat fragmentation on regional bird populations and the need for community involvement in conservation; lessons from the Western Australian wheatbelt. *In* Status and conservation of birds and their habitats in Queensland ed by C. Catterall, P. Driscoll, K. Hulsman and A. Taplin. Queensland Ornithological Society Inc.: Brisbane.

Saunders, D. A. and Curry, P. J., 1990. The impact of agricultural and pastoral industries on birds in the southern half of Western Australia: past and future. *Proc. Ecol. Soc. Aust.* **16**: 303–21.

Saunders, D. A. and de Rebeira, C. P., 1991. Values of corridors to avian populations in a fragmented landscape. Pp. 221–40 *in* Nature conservation 2: the role of corridors ed by D. A. Saunders and R. J. Hobbs. Surrey Beatty & Sons: Chipping Norton, NSW.

Schodde, R., 1981. Bird communities of the Australian mallee: composition, derivation, distribution, structure and seasonal cycles. Pp. 387–416 *in* Mediterranean-type shrublands ed by F. di Castri, D. W. Goodall and R. L. Specht. Elsevier: Amsterdam.

Scougall, A., Majer, J. D. and Hobbs, R., 1993. Edge effects in grazed and ungrazed Western Australian wheatbelt remnants in relation to ecosystem function. In press *in* Nature conservation 3: reconstruction of fragmented ecosystems, global and regional perspectives ed by D. A. Saunders, R. J. Hobbs and P. R. Ehrlich. Surrey Beatty & Sons: Chipping Norton, NSW.

Specht, R. L., 1972. The vegetation of South Australia. Government Printer: Adelaide.

Springett, B. P., 1976. The effect of planting *Pinus pinaster* Ait. on populations of soil microarthropods and on litter decomposition at Gnangara, Western Australia. *Aust. J. Ecol.* **1**: 83–87.

Stoutamire, W. P., 1983. Orchids and wasps. *SWANS* **13**: 8–9.

Strahan, R. E., 1983. The Australian Museum complete book of Australian mammals. Angus and Robertson: London.

Walker, B., 1992. Biodiversity and ecological redundancy. *Conserv. Biol.* **6**: 18–23.

CHAPTER 8

Ecosystem function of biodiversity: experimentation, long-term observations and modelling

S. Lavorel[1] and I. R. Noble[1]

Studying ecosystem function of biodiversity in mediterranean ecosystems involves a certain number of difficulties related to the large spatial and temporal scales concerned, impracticability of manipulative experiments, and complexity of the phenomena. A general methodological framework for integrating long-term observations, experimentation, and modelling is proposed. Difficulties and alternative methods are discussed at each stage. Two illustrative examples of the application of the framework are presented: firstly, the study of the relationship between species diversity and system stability, using and connecting approaches and results from studies on garrigue ecosystems from southern France; secondly, designing a study to investigate the question of species redundancy in the case of Mediterranean tree-grass ecosystems.

Introduction

The importance of biodiversity in ecosystem functioning and its role in buffering systems in the face of climatic uncertainty and land use change have been much discussed but are still little understood. Recent recommendations by SCOPE/IUBS for research on the function of biodiversity have identified the need to set up projects of "long-term ecological research" and "experimentation on ecosystems" (Di Castri and Younès 1990). Because spatial and temporal scales accessible to either experimentation or observations are restricted, and because of the importance and role of variability, modelling studies are needed to both refine our

[1]Ecosystem Dynamics Group, Research School of Biological Sciences, Australian National University, GPO Box 475, Canberra, ACT, Australia 2601.

Pages 149–67 *in* BIODIVERSITY OF MEDITERRANEAN ECOSYSTEMS IN AUSTRALIA ed by Richard J. Hobbs. Surrey Beatty & Sons, 1992.

understanding of ecosystem properties and make predictions about the response of these properties to environmental changes.

It is now widely recognized that co-ordinated projects that successfully link the three approaches of experimentation, long-term observations, and modelling are needed. In this chapter we propose a general methodological framework for the study of ecosystem function of biodiversity. Two examples of emergent properties of ecosystems relevant to the functional role of biodiversity are then presented. Firstly, the relationship between stability and diversity is illustrated with an example from garrigue ecosystems of the French Mediterranean region. Secondly, the general framework is applied to design a study of redundancy in ecosystems, using Mediterranean tree-grass ecosystems.

A general framework

A general framework for approaching the study of ecosystem function in variable environments which vary over long time frames is shown in Figure 1. The approach begins with a hypothesis. A general knowledge of the system, the theoretical context, and information from other systems lead to a conceptual model of the relationships within the system (see, e.g., Eaton and McDonnell 1989), and to the formulation of specific hypotheses. Initially, long-term studies, focused measurements, and experiments are designed to address these hypotheses in a complementary manner. The quality of the observations and experimental designs are critically important if the conclusions to be drawn from them are to be valid. For example, the design must account for factors such as the high climatic variability which characterises Mediterranean ecosystems. Rigourous designs for experimental and observational programmes are also necessary for valid interpretation of the data using appropriate statistical techniques.

Long-term observations

In the context of high variability and environmental change, long-term observations are essential. Certain transient effects (such as effects of short-term environmental variations or fugitive presence of some species), cumulative effects, and rare catastrophic events, can be identified only in long-term studies (Pickett 1989). Many processes which operate at the ecosystem level, such as nutrient cycling or extinction-migration events, also take place over long time scales (Delcourt *et al.* 1983). Once the temporal scale of the processes of interest has been identified, long-term studies should

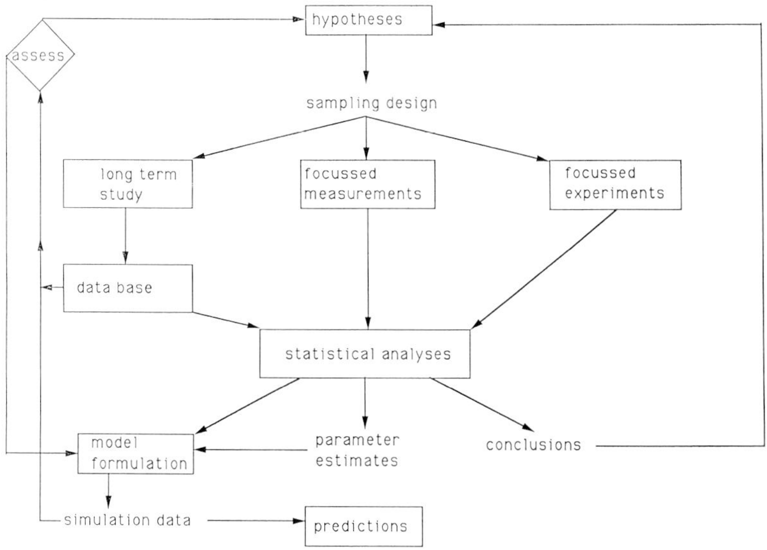

Fig. 1. A general framework for integrated studies in ecology.

be designed so as to cover several dynamic cycles of the system if possible (Eaton and McDonnell 1989).

Unfortunately, if a question needs to be answered rapidly for management decisions, long-term data are rarely readily available. Setting up long-term projects involves a number of practical difficulties. Caraco and Lovett (1989) reviewed various approaches to studying long-term ecological phenomena. Four classes of approaches were distinguished: direct long-term studies, both purely observational and manipulative; retrospective studies using natural (e.g., sediment or ice cores, tree rings) and human (e.g., old data, photos, media accounts) records; modelling; and substitution of space for time (denoted SFT) which assumes that spatial and temporal variation are equivalent. In that case, data on the histories of the sites assumed as elements of a chronosequence are essential. SFT is especially valued as a tool to seek general or qualitative trends, and to generate new hypotheses (e.g., Pickett 1989).

In Australian mediterranean ecosystems, direct or indirect long-term studies need to be designed to compare human-disturbed ecosystems (e.g., agricultural) with adjacent natural remnants. Comparisons between fragmented and unfragmented landscapes would be very useful too, but opportunities of undisturbed habitats are quite limited in the Australian mediterranean region. Also, sets of study sites with different levels of diversity due to historical factors (for example, natural remnants within a same area) could be useful to assess directly the role of the level of diversity. The sets of monitoring sites should cover the precipitation gradient within the Australian mediterranean region (see Hobbs, Chapter 1).

The feasibility and value of long-term studies and their substitutes will depend upon the establishment of networks and organizations responsible for the collection and integration of information from various sources into a data base. Caraco and Lovett (1989) emphasized the difficulties (if not the impossibility) and risks in attempting to formulate a general protocol for long-term studies. Instead, the design should be thought through carefully for each particular question and biological system.

Focused measurements

Accurate measurements of specific parameters will be required for testing particular hypotheses, or to implement simulation models. For these purposes, design of specific measurements are necessary, either at a few points in time and space, or repeatedly over the time of the study. The estimation of some parameters, in particular ecophysiological parameters, requires techniques and precision that are not sustainable in a long-term study. For measurements at small temporal scales (e.g., daily water budget), continuous monitoring over long periods may not be required, and instead measurements over short periods may provide sufficient estimates for calibration of models.

Focused experiments

Manipulative experiments of natural systems are often the ideal way to test the hypotheses. For instance, modifications of the composition of communities by either removal, addition, or change of relative frequencies of species considered as equivalent with respect to a given function, modifications of flows through the system (e.g., water, nutrient or pollen) or of disturbance regimes (fire in particular) provide direct powerful tests of hypotheses relating ecosystem function to level of biodiversity.

Such experiments are not feasible, however, when attempting to understand or predict events that occur at large temporal or spatial scales, such as the changing patterns of climatic variability predicted by climate change scenarios. There are, however, alternative approaches that can provide at least partial or more focused answers.

- Manipulations of artificial communities by varying both the number of species (by introductions and removals), and the regimes of environmental variability (e.g., water regime or fire frequency) will allow an improved understanding of the relationship between diversity and system function. An example is a simple artificial community of annuals in which basic water or nutrient flows are measured in relation to composition. Microcosms are a particular case of easily manipulated systems that can provide large amounts of detailed information within a short time (e.g., Shugart 1989). Organisms with short generation times and rapid responses from perturbations can provide a large number of observations which may serve as alternatives to long-term data sets. Van Voris *et al.* (1980) used a soil-grass microcosm to investigate the relationship between ecosystem complexity and ecosystem stability by relating time of recovery from chemical disturbance to spectral richness of CO_2 emission by microcosms having different species compositions (see Shugart 1989 for details). However, the relevance of such microcosm results to process at ecosystem scale remains debatable.

- Descriptors of ecosystem function can be monitored using natural experiments provided, for instance, by exceptional droughts, large-scale fires, or biological invasions.

- Human land-use has also been the source of a number of large scale "experiments": forestry and land clearing, agriculture, biological control, fire control, replanting of trees and mine rehabilitation are examples. Comparing water and nutrient cycles in these human-manipulated ecosystems with adjacent natural remnants can be expected to give some insight into the role of diversity in these functions.

Some difficulties associated with the interpretation of the focused experiments need to be emphasized. Adding or removing species from simple systems (e.g., rehabilitation, agricultural situations) may not reflect the outcome of similar manipulations in richer systems with many more interspecific interactions. There is also no reason to assume that species additions and removals are exactly inverse processes, so it is necessary to observe situations of species loss as well as of species gains.

Modelling

Models allow access to multiple spatial and temporal scales, and in particular to very large scales. Models can also provide rapid answers for urgent management problems; for instance, how should a landscape be designed so as to conserve functional properties of natural remnants within an agricultural matrix?

Models can be developed in several different ways. A general knowledge of the system and processes operating within it can be used to design a compartmental model (a "box and arrow" type model) for the functions of interest. Such a model can be improved by applying a hierarchical approach identifying components of the levels of organization and separating within and between-level processes (Shugart and Urban 1988).

On the other hand, the analyses of the data obtained through any of the long-term approaches or experiments can lead to a statistical model; for instance, a linear model relating function to ecosystem and environmental parameters. Linear models of water flow based on climatic parameters, soil properties, and abundances of different classes of species have been obtained for Mediterranean tree-grass ecosystems (Shachori *et al.* 1967; Rambal 1987; Joffre and Rambal, in press).

Based on these two types of "primitive" models, and using conclusions from focused mechanistic experiments, more detailed simulation models can be built. The model is developed using parameters estimated from both the data-base and the sets of specific measurements. These simulations are then used to test the model against the data-base and the experimental results. The model's performance in handling exceptional events (e.g., extreme drought, large fires) is also valuable in assessing its relevance. The model can be progressively improved by a process of successive modifications. Sensitivity analyses can be used to indicate the relative roles of different parameters or processes in the model, and also to unravel interactions between parameters or sets of parameters (see Beck 1983).

Simulation results help reassess hypotheses and can lead to the design of new experiments or measurement procedures, either to improve the accuracy of the estimates of parameters to which the model appears to be sensitive, or to answer new questions generated from model results. Simulations also provide a useful tool for predicting the outcomes of management scenarios.

Present and projected environmental changes occur much faster than those experienced throughout evolutionary history. There is also a degree of uncertainty of the precise nature of the changes (Saunders and Hobbs, Chapter 4). Hence models need to be sufficiently general and versatile to incorporate additional data, derived from new observations, long-term studies and exceptional events, as they become available.

General comments and final integration

The scheme presented is very general and needs to be adapted for each particular study. Connections between different approaches are fuzzier than presented in Figure 1. In particular, returns to the hypotheses may occur at any stage, and new experiments and measurements can be undertaken as necessary.

We emphasize the importance of appropriate experimental designs and statistical analyses (see, for example, the discussion by Parker 1989), and hence the need for close interaction with biomathematicians. In particular, methods need to be developed to integrate data collected with different methods, by different operators, and not necessarily from studies specifically designed for the particular question being examined. Integration and extrapolation require a certain number of precautions: clearly defined assumptions and boundary and initial conditions; setting rules for scaling up in both space and time; setting rules for assessing similarity between system components, boundary and initial conditions; and using error estimates (Berkowitz *et al.* 1989).

Example 1. Biodiversity and ecosystem stability

As an example of the general approach described above, we discuss the widely studied and debated question of ecosystem stability and diversity (Fig. 2). The general belief that complex ecosystems should be more stable (MacArthur 1955) predominated for some time in the ecological literature. Elton's argument of the "conventional wisdom" (1958) relied on a list of both theoretical and empirical pieces of evidence to support this view (Table 1). These were drawn from a variety of approaches: long-term data, manipulative experiments of laboratory communities, human and natural large-scale experiments, and mathematical models. Similarly, the controversy that arose from Elton's view resulted in the use of several different approaches to testing the hypothesis (Table 2). As Elton argued, these are lists of mostly

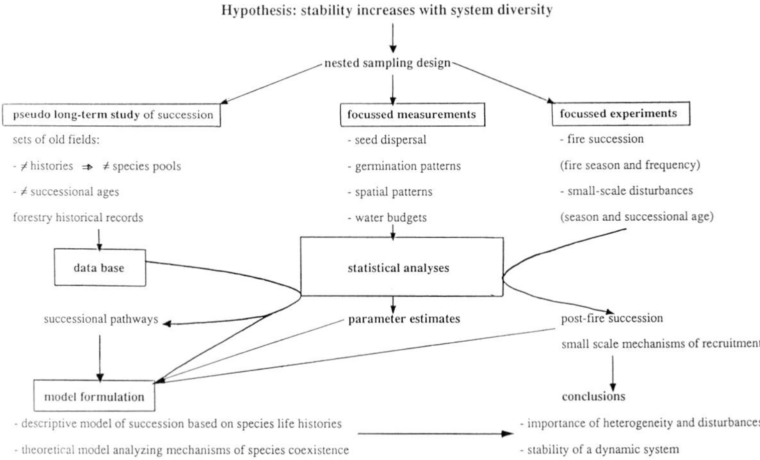

Fig. 2. Methodological framework for the study of the relationship between species diversity and stability in Mediterranean garrigue ecosystems.

Table 1. Arguments of Elton's (1958) "conventional wisdom" (modified from Begon *et al.* 1987).

Approach	Source	Result
Long-term data	Leigh 1975 Wolda 1978	Annual fluctuations of abundances: same level of variability in tropical and temperate or sub-arctic communities
Microcosms	Hairston *et al.* 1968	Laboratory protozoan communities: less distinctions when more diversity
Field and bibliographical data analysis	McNaughton 1978 Briand 1983	Estimates of community connectance: species richness and connectance are negatively correlated
Experimental disturbance	McNaughton 1977	Comparing species-rich and poor communities: perturbations reduce diversity only in species-rich communities
Model	May 1972	Randomly assembled food webs: +complex (e.g., +species) is +unstable
Model	Lawlor 1978 Pimm 1979a	Constrained food webs: +complex is +unstable (but less clear-cut)
Model	De Angelis 1975	Donor-controlled food webs: stability unaffected by complexity
Model	Pimm 1979b	Species deletion in food webs: relationship depending on trophic level
Model	Pimm and Post 1983 Case 1991	Species invasion of communities: more difficult in species-rich communities

Table 2. Approaches to, and evidence for, the issue of diversity and stability.

Approach	Statement
Long-term data	Species-rich tropical communities are not noted for insect outbreaks when compared to their temperate and boreal counterparts
Natural experiments	Islands, which usually possess few species, are more vulnerable to invasion than species-rich continents
	Crop monocultures are peculiarly vulnerable to invasion and destruction by pests
Laboratory communities	Simple artificial communities are prone to extinctions
Mathematical models	Models of interactions between two or few species are inherently unstable

unconnected observations which are consistent with the complexity/stability hypothesis, but which can also be explained by alternative hypotheses (Begon *et al.* 1987), so the issue remains open.

This failure to find an effective answer to the question has at least two main causes:

- confusion over the concepts: stability has never been defined clearly and uniformly for these studies, and no clear distinction has been made between diversity and complexity;
- the lack of a co-ordinated approach linking theoretical, observational, and experimental studies.

We propose here to apply the integrated scheme presented above to test the hypothesis that stability increases with diversity. We use the definition of stability as the return of the system to equilibrium after any displacement and the set of properties relating to stable systems that follow (Chesson and Case 1986; Table 3; but see also Majer, Chapter 9). Table 3 also indicates possible methods for testing the four properties.

Unfortunately, there is to our knowledge no Australian example of studies having attempted to integrate the different approaches. We thus use pieces of research on "garrigue" ecosystems from Southern France to connect them into the general scheme. The research was not designed specifically for the question we are tackling here, nor was it originally integrated into such a scheme. It is connected only by the underlying interest to understand the dynamics of the garrigue landscapes.

Table 3. Properties of stable communities and methodologies for their tests.

	Property	Formulation	Tests
P1	Community conservation	In the absence of perturbation the community never loses species	Long-term data, microcosms simulation models
P2	Community recovery	The community can recover from any event driving species to low density	Manipulative and natural experiments, long-term demographic data (statistical models), simulation models
P3	Community assembly	Community build up by immigration of species and eventually reach equilibrium	
P4	Irrelevance of history	Effects of past abundances disappear in the long term	

Garrigue landscapes are mosaics of old fields at different successional stages and of *Quercus ilex* forests. They are characterized by a high spatial heterogeneity at a variety of scales, a high species diversity, and the frequency and variety of disturbances at all scales (see Lavorel *et al.* 1991a,b).

Long-term studies

The structure and dynamics of communities at different successional stages has been monitored both directly through time with permanent plots since 1969, and indirectly using fields at different successional stages with similar histories (Escarré *et al.* 1983; Escarré and Debussche, in prep.). Hence, a data set is available with communities having different levels of diversity in relation to successional age and disturbance history. Data analyses have yielded schemes of successional pathways (Escarré *et al.* 1983), and helped in understanding community assembly and recovery.

Forestry and rural planning data for several centuries (since the XIVth century in some townships) have also been useful to reconstruct the dynamics of forest cover and landscape patterns. These data indicate that in spite of dramatic changes in land use patterns (with, for instance, increases and decreases of forested area), garrigue systems are remarkably resilient and there has been no significant decrease in diversity at the landscape scale (Debussche *et al.* 1987; Debussche and Lepart 1990).

Focused measurements

These have been made to quantify spatial patterns (Lavorel *et al.* 1991a), seed dispersal (e.g., Olivieri *et al.* 1983; Acherar *et al.* 1984; Debussche *et al.* 1985), germination patterns (Lavorel *et al.*, in press), and water budgets (Rambal 1987; Rambal *et al.* 1984; Debussche *et al.* 1987).

Manipulative experiments

Community recovery was studied after controlled fires in a combination of frequency and season regimes in *Quercus coccifera* garrigues. These communities appeared exceptionally stable in relation to life form diversity of component species (regenerating vegetatively *vs* sexually via a seed bank) under virtually any fire regime (Trabaud and Lepart 1980, 1981; Malanson and Trabaud 1987). The conclusions were that, in these communities, community recovery and irrelevance of history resulted from the diversity in the attributes of the component species.

Community recovery and community assembly in relation to multiscale heterogeneity and small-scale disturbances was studied experimentally by Lavorel *et al.* (1991b). Disturbances were artificially generated in different seasons in old fields of different ages. Results suggested that diversity at the landscape scale is maintained through the interaction between the existence of a variety of guilds of species with equivalent, but seasonally specific, recruitment niches and disturbances that vary in space and time.

Modelling

A model of community dynamics (Prado 1987) showed that successional pathways and spatial patterns consistent with the long-term data can be simulated based only on the initial floristic composition (Egler 1954) and biological properties of the component species. Parameters were directly incorporated from measurements and general knowledge of the system. This model could be used to examine stability with different degrees of diversity, but this has not yet been done.

Using the above results on relationships between community dynamics and stability, and species recruitment biology, a theoretical model of community dynamics in mosaic landscapes with local disturbances (Lavorel *et al.* 1991c) has been developed to investigate stability of species assemblages with different sets of life histories and habitat patterns under different disturbance regimes. Simulations have examined the questions of recovery from disturbance and irrelevance of history, and indicate that species coexistence is favoured by diversity in life histories.

Conclusion

Linking the results from different approaches suggests that the stability of the whole system is linked to its diversity: the existence of groups of species able to respond to different disturbances and the existence of several equivalent species within these groups. Further, they suggest that the diversity of the landscape itself is important, with a mosaic of patches with different levels of diversity and species compositions. However, this *a posteriori* connection of more or less independent studies cannot fully test the initial hypotheses and more specific experiments would need to be designed. The existing models would also need to be used in specific simulations comparing the stability of species assemblages with different levels of diversity, both in terms of number of "recruitment guilds", and of diversity within these groups.

Example 2. Redundancy in ecological systems

The notion of species functional redundancy, that is the idea that "at large scales, biodiversity is not important in carbon, nutrient, and water balance due to internal redundancies and compensations" (Solbrig 1991a), has received considerable attention recently (Simpson 1988; Di Castri and Younès 1990; Di Castri 1991; Hobbs, Chapter 1). Redundancy can be said to exist within a group of species if deleting one of them from the system does not modify the behaviour of the system as a whole. Redundancy is a very ambiguous notion, since it is strongly and inevitably related to the particular system and function under consideration. The relevance of this notion to management of biodiversity is quite debatable. However, if the goals of conservation are to maintain ecosystem properties such as water, nutrient, or energy fluxes, or general food web structure, then groups of species may be considered as equivalent and hence replaceable regarding given ecosystem functions. Relevant questions are then for example: Is a certain level of redundancy necessary for ecosystem persistence (see Main, Chapter 5)?; Is redundancy an artefact of ignoring the role of particular species in carrying out particular ecosystem functions under specific sets of environmental conditions?

We would like to examine how the integrated methodological framework could be used to diagnose redundancy regarding water balance in mediterranean climate tree-grass ecosystems. The application of an adapted version of the general scheme to this case study is presented in Figure 3. A stepwise "hierarchical" approach seems necessary. Firstly, the number of functional groups present and the number needed to maintain ecosystem processes while retaining the capacity of response to possible changes are defined. Secondly, the diversity within these groups is quantified, and estimates are made of the numbers of "equivalents" needed in each group.

It is believed that water-limited natural vegetation equilibrates towards a tree-grass composition producing minimum water stress of individual components and maximum biomass (Eagleson 1982; Eagleson and Segarra 1985). Studies in Spanish dehesas (Joffre *et al.* 1987) and in Californian oak savannahs (Holland 1973) have indicated an important functional diversity within the grass stratum, with coexistence of annual and perennial species with very different life histories. Setting aside the question of tree-grass proportions, we propose a study design to identify functional types within the grass layer, and to test whether the concept of redundancy can apply in such a context. More precisely, the question becomes: can we identify groups of functional equivalents that are responsible for the maintenance of the hydrological dynamics in the system, regardless of the specific identities?

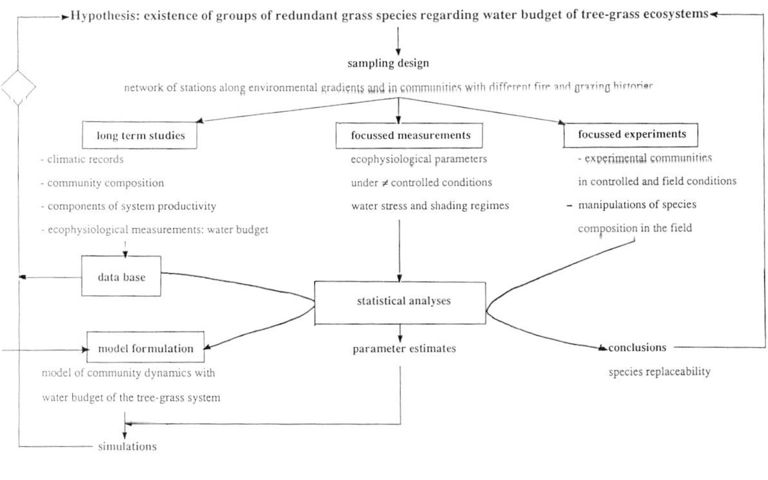

Fig. 3. Methodological framework for the study of species functional redundancy in Mediterranean tree-grass ecosystems.

Long-term data

A network of stations needs to be set up along environmental gradients (e.g., degree of aridity in the case of Western Australia; see Hobbs, Chapter 1) and including sites with different histories (land use, fire). These stations would monitor:

- climatic data (temperature, precipitation, insolation); data from surrounding stations (microstations on farms, airports . . .) can also be included into the data base;
- community composition: tree density, composition of the grass stratum;
- soil water storage (1) and its temporal variations (monthly and annual water balance);
- water use by trees (2), estimated from measures of actual evapotranspiration and of conductivity (Sisson *et al.* 1980); water use by grasses is then deduced by the difference between (1) and (2);
- components of system productivity: annual biomass production of grasses, root/shoot ratios, etc;
- more specific ecophysiological measurements on target grass species, estimating monthly water uptake, evapotranspiration and biomass production.

Analyses of these data will form the basis of a functional model of the community water use in relation to the abundances of component species. They may also lead to a gross classification of species into functional groups. Different groupings can be developed under different criteria in order to estimate the degree of overlap between different groupings.

Experimental approach

The experimental component would have as its goals:

a. Parameter estimation: greenhouse and growth chamber measurements of ecophysiological parameters for the different species under different regimes of water stress (inspired from natural weather pattern, emphasizing the rainfall unpredictability factor) and shading (mimicking shading by trees). Some parameters of the dynamics of water use by trees will also need to be estimated in order to implement the ecosystem model.

b. Refinement of the understanding of mechanisms:

   - in the field, and for a sample of representative stations, monitoring of the above ecophysiological parameters after manipulation of densities or the presence-absence of different species;

- in controlled greenhouse or garden experiments measuring water use efficiency for experimental communities manipulating both frequencies of different functional groups, and diversity within these groups, and comparing results to the measures in natural communities.

Modelling

A model derived from MUSE (Solbrig 1991b) could be developed to examine the role of functional diversity and redundancy in the water balance of tree-grass ecosystem. Such a spatially-explicit model deals with the dynamics of tree and grass components and their interactions (e.g., shading, root competition). The model can simulate either individual species or different functional types, and could use the groupings obtained from the data base, or from some of the focused experiments. The model is constructed as a shell in which different biological functions can be handled and changed separately, hence it is possible to implement a more detailed ecophysiological submodel of water use, and to calculate daily, seasonal, and annual water budgets, using the same sets of parameters and variables as for field measurements.

Simulations using the parameter sets measured in the field will allow us to:

- test whether using the species groupings yields the same water dynamics as in the field;
- analyse sensitivity to different parameters and assess whether the natural range of variation of these parameters between "equivalent species" is compatible with the assumption that interchanging species does not affect system function;
- modify the number and composition (i.e., how many and which species) of the functional types and examine how this modifies the system function;
- modify tree cover and/or climate and examine whether the functional types defined for the grasses are still valid.

Conclusion

The development of integrated and co-ordinated multidisciplinary approaches is crucial for successfully tackling the question of diversity and ecosystem function, especially at scales not directly accessible to humans. This issue is of diversity and ecosystem function is urgent and it is essential for scientists to be able to provide managers with predictions and assessments of the impacts of management actions. An important element for the success of an integrated approach is its implementation in the original design of a research programme, as opposed to patching together of pieces of on-going or past research. Also, it is crucial that integration of available data and co-ordination of projects be organized at the regional scale. We suggest that such criteria be taken into account in the assessment of scientific projects and funding allocation.

References

Acherar, M., Lepart, J. and Debussche, M., 1984. La colonisation des friches par le Pin d'Alep (*Pinus halepensis,* Miller) en Languedoc méditerranéen. *Acta Ecol. Oecol. Plant.* **19:** 179–89.

Beck, M. B., 1983. Sensitivity analysis, calibration and validation. Pp. 425–67 *in* Mathematical modelling of water quality: streams, lakes and reservoirs ed by G. T. Orlob. International Series on Applied Systems Analysis Volume 12. Wiley and Son: Chichester, England.

Begon, M., Harper, J. L. and Townsend, C. R., 1987. Ecology: individuals, populations and communities. Blackwell Scientific Publications: Oxford.

Berkowitz, A. R, Kolasa, J., Peters, R. H. and Pickett, S. T. A., 1989. How far in space and time can results from a single long-term study be extrapolated? Pp. 192–98 *in* Long-term studies in ecology — Approaches and alternatives ed by E. G. Likens. Springer-Verlag: New York.

Briand, F., 1983. Environmental control of food web structure. *Ecology* **64:** 253–63.

Caraco, N. M. and Lovett, G. M., 1989. How can the various approaches to studying long-term ecological phenomena be integrated to maximize understanding? Pp. 187–89 *in* Long-term studies in ecology — Approaches and alternatives ed by E. G. Likens. Springer-Verlag: New York.

Case, T. J., 1991. Invasion resistance, species build-up and community collapse in metapopulation models with interspecific competition. *Biol. J. Linn. Soc.* **42:** 239–66.

Chesson, P. L. and Case, T. J., 1986. Overview: nonequilibrium community theories, chance, variability, history and coexistence. Pp. 229–39 *in* Community ecology ed by J. Diamond and T. J. Case. Harper and Row: New York.

De Angelis, D. L., 1975. Stability and connectance in food web models. *Ecology* **56:** 238–43.

Debussche, M., Lepart, J. and Molina, J., 1985. La dissémination des plantes à fruits charnus par les oiseaux: rôle de la structure de la végétation et impact sur la succession en région méditerranéenne. *Acta Ecol. Oecol. Gener.* **6:** 65–80.

Debussche, M., Rambal, S. and Lepart, J., 1987. Les changements de l'occupation des terres en région méditerranéenne humide: évaluation des conséquences hydrologiques. *Acta Ecol. Oecol. Applic.* **8:** 317–32.

Debussche, M. and Lepart, J., 1990. Human impact on landscape patterning: Mediterranean examples. Pp. 76–106 *in* Landscape boundaries: consequences for biotic diversity and ecological flows ed by F. di Castri and A. J. Hansen. Springer-Verlag: New York.

Delcourt H. R., Delcourt P. A. and Webb T. III., 1983. Dynamic plant ecology: the spectrum of vegetational change in space and time. *Quatern. Sci. Rev.* **1**: 153–75.

di Castri, F., 1991. Ecosystem evolution and global change. Pp. 189–218 *in* Perspectives in biological complexity ed by O. T. Solbrig and G. Nicolis. IUBS: Paris.

di Castri, F. and Younès, T. (eds), 1990. Ecosystem function of biodiversity. *Biology International Special Issue 22.*

Eagleson, P. S., 1982. Ecological optimality in water-limited natural soil-vegetation systems, 1. Theory and hypotheses. *Water Resource Research* **18**: 325–40.

Eagleson, P. S. and Segarra, R. I., 1985. Water limited equilibrium of savanna vegetation systems. *Water Resource Research* **21**: 1483–93.

Eaton, J. S. and McDonnell, M. J., 1989. What are the difficulties in establishing and interpreting the results from a long-term manipulation? Pp. 190–92 *in* Long-term studies in ecology — Approaches and alternatives ed by E. G. Likens. Springer-Verlag: New York.

Egler, F. E., 1954. Vegetation science concepts. I. Initial floristic composition — a factor in old-field vegetation development. *Vegetatio* **4**: 412–17.

Elton, C. S., 1958. The ecology of invasions by animals and plants. Methuen, London.

Escarré, J., Houssard, C., Debussche, M. and Lepart, J., 1983. Evolution de la végétation et du sol après abandon cultural en région méditerranéenne: étude de la succession dans les garrigues du Montpelliérais (France). *Acta Ecol. Oecol. Plant.* **18**: 221–39.

Hairston, N. D., Allan, J. D., Futuyama, D. J., Howell, J., Lubin, L. B., Mathias, J. and Vandermeer, J. H., 1968. The relationship between species diversity and stability: an experimental approach with protozoa and bacteria. *Ecology* **49**: 1091–01.

Holland, V. L., 1973. A study of vegetation and soils under blue oak compared to adjacent open grassland. Ph.D. Thesis, University of California, Berkeley, USA.

Joffre, R., Leiva Morales, M. J., Rambal, S. and Fernandez Ales, R., 1987. Dynamique racinaire et extraction de l'eau du sol par des graminées annuelles et pérennes méditerranéennes. *Acta Ecol. Oecol. Plant.* **8**: 181–94.

Joffre, R. and Rambal, S., in press. How do trees modify the water balance of Mediterranean rangelands? *Ecology.*

Lavorel, S., Lebreton, J. D., Debussche, M. and Lepart, J., 1991a. Nested spatial patterns in seed bank and vegetation of Mediterranean old fields. *J. Veg. Sci.* **2**: 367–76.

Lavorel, S., Lepart, J., Debussche, M. and Lebreton, J. D., 1991b. Small scale disturbances and the maintenance of diversity in Mediterranean old fields. Pp. 105–61 *in* Spatial structure, disturbances, and the dynamics of coexistence of plant species: from experimentation to modelling. The example of Mediterranean old fields. Thèse de Doctorat, Université des Sciences et Techniques du Languedoc, Montpellier, France.

Lavorel, S., Gardner, R. H. and O'Neill, R. V., 1991c. Coexistence in patchy landscapes of annual plant species with different spatio-temporal dispersal strategies. Pp. 207–37 *in* Spatial structure, disturbances, and the dynamics of coexistence of plant species: from experimentation to modelling. The example of Mediterranean old fields. Thèse de Doctorat, Université des Sciences et Techniques du Languedoc, Montpellier, France.

Lavorel, S., Debussche, M., Lebreton, J. D. and Lepart, J., in press. Seasonal patterns in the seed bank of Mediterranean old fields. *Oikos*.

Lawlor, L. R., 1978. A comment on randomly constructed ecosystem models. *Amer. Nat.* **112:** 445–47.

Leigh, E., 1975. Population fluctuations and community structure. Pp. 67–88 *in* Unifying concepts in ecology ed by W. H. van Dobben and R. H. Lowe McDonnell. Junk: The Hague.

MacArthur, R. H., 1955. Fluctuations of animal populations and a measure of community stability. *Ecology* **36:** 533–36.

McNaughton, S. J., 1977. Diversity and stability of ecological communities: a comment on the role of empiricism in ecology. *Amer. Nat.* **11:** 515–25.

McNaughton, S. J., 1978. Stability and diversity in ecological communities. *Nature* **274:** 251–52.

Malanson, G. P. and Trabaud, L., 1987. Ordination analysis of components of resilience of *Quercus coccifera* garrigue. *Ecology* **68:** 463–72.

May, R. M., 1972. Will a large complex system be stable? *Nature* **238:** 413–14.

Olivieri, I., Swan, M. and Gouyon, P. H., 1983. Reproductive system and colonizing strategies of two species of *Carduus*. *Oecologia* **60:** 114–17.

Parker, G. G., 1989. Are currently available statistical methods adequate for long-term studies. Pp. 199–200 *in* Long-term studies in ecology — Approaches and alternatives ed by E. G. Likens. Springer-Verlag: New York.

Pickett, S. T. A, 1989. Space for time substitution as an alternative for long-term studies. Pp. 112–35 *in* Long-term studies in ecology — Approaches and alternatives ed by E. G. Likens. Springer-Verlag: New York.

Pimm, S. L., 1979a. Complexity and stability: another look at MacArthur's original hypothesis. *Oikos* **33:** 351–57.

Pimm, S. L., 1979b. The structure of food webs. *Theor. Popn. Biol.* **16:** 144–58.

Pimm, S. L. and Post, W. M., 1983. Community assembly and ecosystem stability. *Math. Biosci.* **64:** 169–92.

Prado, C., 1987. Un modèle de succession végétale: rôle des traits biologiques et des contraintes spatiales. Thèse de Doctorat, Université Paris VI, France.

Rambal, S., 1987. Evolution de l'occupation des terres et ressources en eau en région méditerranéenne karstique. *J. Hydrol.* **93:** 339–57.

Rambal, S., Ibrahim, M. and Rapp, M., 1984. Variabilité spatiale des variations du stock d'eau du sol sous forêt. Application à l'optimisation d'un dispositif de mesure du bilan hydrique. *Catena* **11:** 177–86.

Shachori, A. Y., Rosenzweig, D. and Poljakoff-Mayber, A., 1967. Effect of mediterranean vegetation on the moisture regime. Pp. 291–311 *in* Forest hydrology ed by W. E. Sopper and H. W. Lull. Pergamon: Oxford.

Shugart, H. H., 1989. The role of ecological models in long-term ecological studies. Pp. 91–111 *in* Long-term studies in ecology — Approaches and alternatives ed by E. G. Likens. Springer-Verlag: New York.

Shugart, H. H. and Urban, D. L., 1988. Scale, synthesis, and ecosystem dynamics. Pp. 279–89 *in* Concepts of ecosystem ecology ed by L. R. Pomeroy and J. J. Alberts. Springer-Verlag: New York.

Sisson, J. B., Ferguson, A. H. and Van Genutchen, M. Th., 1980. Simple methods for predicting drainage from field plots. *Soil Soc. Am. J.* **44**: 1147–52.

Solbrig, O. T., 1991a. From genes to ecosystems: a research agenda for biodiversity. IUBS: Paris.

Solbrig, O. T., 1991b. Savanna modelling for global change. *Biology International, Special Issue* **24**.

Simpson, B. B., 1988. Biodiversity in the context of ecosystem structure and function. *Biology International* **17**: 15–17.

Trabaud, L. and Lepart, J., 1980. Diversity and stability in garrigue ecosystems after fire. *Vegetatio* **43**: 49–57.

Trabaud, L. and Lepart, J., 1981. Changes in floristic composition in a *Quercus coccifera* L. garrigue in relation to different fire regimes. *Vegetatio* **46**: 105–16.

Van Voris, P., O'Neill, R. V., Emmanuel, W. R. and Shugart, H. H., 1980. Functional complexity and ecosystem stability. *Ecology* **61**: 1352–60.

Wolda, H., 1978. Fluctuations in abundance of tropical insects. *Amer. Nat.* **112**: 1017–45.

PLATE 5

ABOVE: A recently-burned area of heathland on Durokoppin Nature Reserve, with a carpet of *Glischrocaryon aureum*, which declines in abundance as heath shrubs regrow. (Photo: R. J. Hobbs).

BELOW: Heathland in the Fitzgerald River National Park, with *Hakea victoria* in the foreground. The park is the largest unfragmented area of natural vegetation in the agricultural area and contains more vascular plant species than the whole of the British Isles. (Photo: R. J. Hobbs).

CHAPTER 9

Maintaining and restoring diversity

J. D. Majer[1]

This chapter discusses the vulnerability of mediterranean ecosystems within Australia to disturbance and the ease with which they may be restored once they have been altered. In the absence of comparative studies on disturbance, the ecosystems in both hot (type Csa) and relatively cooler (type Csb) mediterranean areas are looked at in terms of the phenophases of their various components. The possible influence of episodic disturbances in spring and autumn on these ecosystems is then postulated.

Comparative data are available on the resilience of ecosystems in mediterranean Australia. Some studies suggest that mediterranean regions are more resilient than most non-mediterranean regions, although there is no evidence for difference in resilience between the different mediterranean regions of the world. By contrast, data on the return of ants in rehabilitated minesites in Australia suggest that, for this taxon at least, mediterranean regions exhibit lower resilience than the non-mediterranean regions which were investigated.

Although ecological functions can be reduced by disturbance of the community, examples are given where decomposition can continue in areas where the decomposer-associated fauna is reduced and where decomposition can be restored in rehabilitated minesites when the fauna (and microflora) returns. This chapter concludes that more comparative studies, designed to relate the inertia and resilience of mediterranean ecosystems to other regions, should be initiated.

Introduction

In view of the fact that this information was prepared for the Australian "Workshop on the Function of Biodiversity in Mediterranean Ecosystems", I have taken the approach of providing some data and thoughts on what is actually happening in the mediterranean parts of Australia, rather than trying to derive

[1]School of Environmental Biology, Curtin University of Technology, PO Box U1987, Perth, Western Australia, Australia 6001.

Pages 169–91 *in* BIODIVERSITY OF MEDITERRANEAN ECOSYSTEMS IN AUSTRALIA ed by Richard J. Hobbs. Surrey Beatty & Sons, 1992.

some general principles on the maintenance and restoration of diversity. Certain aspects of these phenomena have already been addressed in the publications "Resilience in Mediterranean-type Ecosystems" (Dell *et al.* 1986) and "Australian Ecosystems: 200 Years of Utilization, Degradation and Reconstruction" (Saunders *et al.* 1990) and further attempts were made to extract general principles for mediterranean ecosystems at the SCOPE/IUBS-sponsored conference on "Biotic Diversity and Ecosystem Function" which was held in Cape Town in September 1992.

In order to undertake this review, it is first necessary to define the areas of Australia which will be considered. For the purposes of this discussion I will adhere to those areas of Western Australia, South Australia and Victoria which possess Köppen's climatic zones Csa (hot summer) and Csb (cooler summer). These areas have been described by Hobbs (Chapter 1).

Conservation status of mediterranean regions of Australia

Before discussing the maintenance of biodiversity in these regions, it is important to document the conservation status of the various physiognomic units. In Western Australia, Beard and Sprenger (1984) have measured the total area of each physiognomic unit and also the percentage of that which has been cleared or which is in private hands and probably subject to clearing in the near future. Some 65 per cent of the south-west botanical province has been cleared, although this proportion is much higher for certain units (Table 1; see also Saunders and Hobbs, Chapter 4). Units 15 and 30 appear to have virtually disappeared (there are remnants of unit 30 on private land) while units 17, 31, 7, 2, 4 and 13 have been cleared of over 90 per cent of their area and are thus perhaps under threat. The main threats to these physiognomic units, which are situated in the most densely populated part of the State, are agricultural, urban and industrial use. There are, however, more benign threats to these units such as weed invasion, inappropriate fire regimes and many others; a list of threats to plant species diversity is provided by both Hobbs (Chapter 1) and Lamont (Chapter 6).

Comparable figures for the mediterranean region of south-east Australia are not so readily available. Some information is available in Specht *et al.* (1974), while Laut *et al.* (1977) comment on some of the human activities in the various environmental provinces. It appears that, because of the higher human population in eastern than western Australia and also because of the

Table 1. List of physiognomic units in the south-west botanical province showing their total original area, their percentage contribution to the area of the province and the percentage of each unit which has been cleared (adapted from Beard and Sprenger 1984).

Unit No.	Physiognomic unit	Floristic	Area km^2	% of total	% cleared
1	Tall Forest	*Eucalyptus diversicolor, E. marginata, E. calophylla*	4004	1.29	14
2	Tall Woodland	*E. gomphocephala*	69	0.02	94
3	Forest	*E. marginata*	31 802	10.26	44
4	Woodland	*E. calophylla-E. wandoo*	16 697	5.39	94
5	Woodland	*E. astringens-E. accedens*	958	0.31	72
6	Woodland	*E. gomphocephala*±*E. marginata*	1 039	0.34	90
7	Woodland	*E. loxophleba, E. salmonophloia, E. wandoo*	41 126	13.27	97
8	Woodland	*E. salmonophloia-E. salubris*	17 580	5.67	78
9	Woodland	*E. le souefii-E. transcontinentalis-E. oleosa*	950	0.31	0
10	Woodland	*E. oleosa-E. flocktoniae*	1 575	0.51	0
13	Open Woodland	Mainly *E. wandoo*	221	0.07	94
14	Low Forest	*E. marginata*	964	0.31	82
15	Low Forest	*Callitris preissii*	29	0.01	100
16	Low Forest	*E. platypus* or *E. cornuta-E. lehmannii*	113	0.04	2
17	Low Forest	*Acacia rostellifera*	327	0.10	98
22	Low Woodland	*Agonis flexuosa*	242	0.08	74
23	Low Woodland	*Banksia* spp.±*E. marginata*	6 229	2.01	61
25	Low Woodland	*Allocasuarina huegeliana-E. loxophleba*	88	0.03	88
26	Low Woodland	*E. marginata-Allocasuarina fraserana*	901	0.29	41
27	Low Woodland	*Melaleuca* spp.	926	0.30	49
30	Low Woodland +scattered emergent trees	*Banksia* spp.+*E. marginata*	680	0.22	100
31	Thicket with ditto	*Allocasuarina campestris*+ *E. wandoo* or *Melaleuca uncinata* +*E. loxophleba*	441	0.14	98
35	Scrub+ scattered trees	*Acacia* spp.+ *Eucalyptus* and/or *Callitris*	5 354	1.73	83
36	Thicket	*Acacia-Allocasuarina-Melaleuca*	35 496	11.46	80
37	Thicket	*Melaleuca rhaphiophylla*	143	0.05	74
38	Thicket	*Dryandra* spp. and/or *Eucalyptus* spp.	497	0.16	18
40	Scrub	*Acacia* spp.	1 109	0.36	78
41	Scrub	*Melaleuca* spp.	76	0.02	70
42	Scrub	*Agonis flexuosa, Acacia, Eucalyptus*	1 636	0.53	12
44	Sparse scrub	*Acacia* spp.	121	0.04	66
45	Mallee	*Eucalyptus* spp.	46 385	14.97	55

Table 1. — continued

Unit No.	Physiognomic unit	Floristic	Area km^2	% of total	% cleared
131	Mallee and woodland patches	*Eucalyptus* spp.	22 219	7.17	40
47	Mallee-heath	*E. tetragona*+mixed Proteaceae-Myrtaceae	17 114	5.52	69
47a	Tree-heath	Heterogeneous	3 626	1.17	0
48	Scrub heath	Heterogeneous, Proteaceae-Myrtaceae	35 070	11.32	61
49	Heath	Heterogeneous, Proteaceae-Myrtaceae	2 554	0.82	41
51	Sedgeland	Cyperaceae, Restionaceae	499	0.16	48
119	Succulent steppe, thickly wooded	*Atriplex, Maireana+Eucalyptus* spp.	4 486	1.44	87
121	Ditto, lightly wooded	*Atriplex, Maireana+Acacia* spp.	104	0.03	68
124	Ditto, unwooded	Samphire: *Halosarcia*	313	0.10	64
125	Bare areas: Playa lakes		4 896	1.58	30
126	Ditto: Freshwater lakes		123	0.04	54
128	Ditto: Rock outcrops		413	0.13	54
129	Ditto: Drift sand		645	0.21	10
44 Units		Total	309 840	100	65
			12.3% of State		

extent of agriculture, the level of clearing in the mediterranean parts of eastern Australia is greater, and the percentage of plant alliances not represented in conservation areas is less, than in southwestern Australia.

Terminology

Before discussing the maintenance and restoration of biodiversity, it is important to clarify the terminology relating to a community's ability to resist change and its ability to recover following disturbance.

There is a profusion of conflicting terminology relating to this subject so, to avoid confusing the reader, I have adopted that recently put forward by Westman (1986) and have represented it diagrammatically in Figure 1. The vertical axis of Figure 1 could describe the population density of an individual species, the species richness of the entire community or it could be an index which integrates characteristics of both the physical and biotic environment. Prior to disturbance the ecosystem exists at a pre-disturbance level. The ecosystem becomes stressed as a disturbance is initiated. The degree of strain inflicted on the ecosystem depends on both the magnitude and period of stress. Thus the strain on a heathland would be larger during opencut mining than burning.

The disturbance process and the period of recovery may be characterized by two parameters, "inertia" and "resilience". Inertia is the ability of the ecosystem to resist the external stress which is applied to it, while resilience is the degree, manner, and

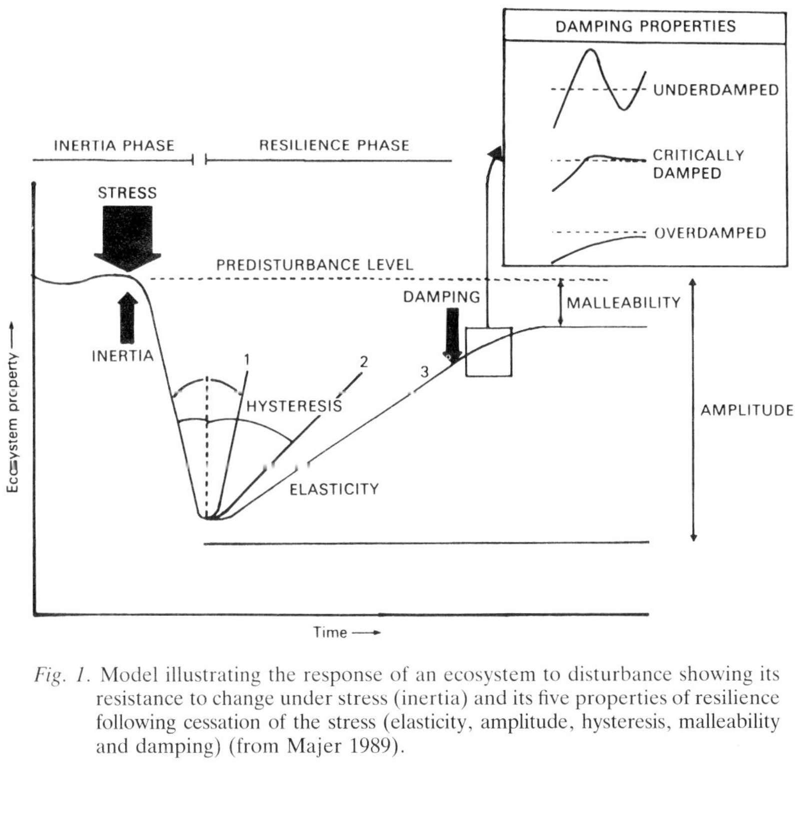

Fig. 1. Model illustrating the response of an ecosystem to disturbance showing its resistance to change under stress (inertia) and its five properties of resilience following cessation of the stress (elasticity, amplitude, hysteresis, malleability and damping) (from Majer 1989).

pace of recovery of the ecosystem to the predisturbance level. Westman (1986) has identified four properties of resilience: "elasticity", "amplitude", "hysteresis" and "malleability" (Fig. 1).

Elasticity is the rate of recovery of an ecosystem property following disturbance. Thus, the steeper the slope of recovery, the more elastic is the ecosystem. The amplitude component of resilience is the threshold level of strain beyond which return to the original state no longer occurs. This is represented by the difference between the predisturbance level and the solid horizontal line on Figure 1. Hysteresis concerns the degree to which the path of change in an ecosystem property under stress differs from the path of succession upon immediate removal of the stress. The malleability of the ecosystem is the difference between the ecosystem's final recovery level and the predisturbance level; the greater the difference, the more malleable is the ecosystem.

Fox and Fox (1986) described a fifth property of resilience: "damping". This is the degree and manner by which a path of restoration is altered by any forces that affect the restoring force. Figure 1 shows the path of recovery of overdamped, critically damped and underdamped ecosystems. The overdamped ecosystem will take a long time to reach equilibrium but will not oscillate. The underdamped ecosystem will approach equilibrium rapidly but will overshoot and oscillate for some period of time, while the critically damped ecosystem will approach equilibrium more slowly, overshoot slightly and then approach equilibrium level. This ecosystem equilibrates most rapidly and is therefore the most resilient of the three examples.

The last-mentioned property assumes the existence of an equilibrium. There is an increasing body of literature which provides evidence that natural ecosystems do not necessarily possess fixed equilibria (De Angelis and Waterhouse 1987). If this is the case, it may be extremely difficult to measure the impact of a stress on biodiversity and also the degree to which this impact persists.

Phenology in mediterranean regions of Australia

In order to understand the vulnerability of mediterranean ecosystems to disturbance, I believe that it is important to consider the phenological events. The annual climatic sequence in mediterranean regions is characterized by marked contrasts. Groves and Hobbs (Chapter 3) have already mentioned that fluctuations in water and temperature, along with scarce nutrients, are the dominant influences on the inertia and resilience of mediterranean plant and animal communities within Australia. In view of the fact that there is generally a gradation in the severity of mediterranean climatic patterns from the cooler and wetter to the hotter and drier areas (Hobbs, Chapter 1), I do not believe that the degree of inertia and resilience is constant throughout these regions. I therefore present data from two sites in Western Australia, one close to Perth and a warmer, drier site at Durokoppin Nature Reserve, near Kellerberrin. These sites have been selected to exhibit the differences in phenologies between cooler and hotter sites and also because reasonable and comparable data sets are available for both areas. Data for the "Perth" site are in fact drawn from a range of sites between Perth and Dwellingup, over a 70 km distance.

Figure 2 shows monthly climatic values for representative years at Dwellingup and for the Kellerberrin area. Both sites exhibit the mediterranean pattern with cool, wet winters and hot dry summers. However, summer temperatures near Kellerberrin are higher than at Dwellingup and rainfall is considerably lower (the high February value for Kellerberrin was exceptional). In addition to being drier, the rainfall pattern is more closely tied to the middle of the year at Kellerberrin than it is at Dwellingup (see also Hobbs, Chapter 1; Saunders and Hobbs, Chapter 4).

The growth of vegetation is intimately tied to water availability. Although no community data are available for the Kellerberrin area, data for Cannington, near Perth (Fig. 3) indicate two main peaks in plant growth, one in spring-early summer and a smaller peak in winter. Stephens (1979) divided the plants of the

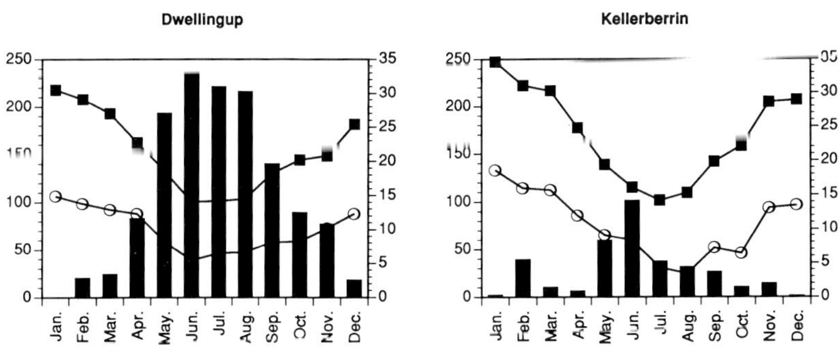

Fig. 2. The monthly rainfall (histogram) and mean monthly temperatures (■–■ = maximum, ○–○ = minimum) at Dwellingup (1981 data from Postle *et al.* 1986) and near Kellerberrin (1986 data from van Schagen 1986).

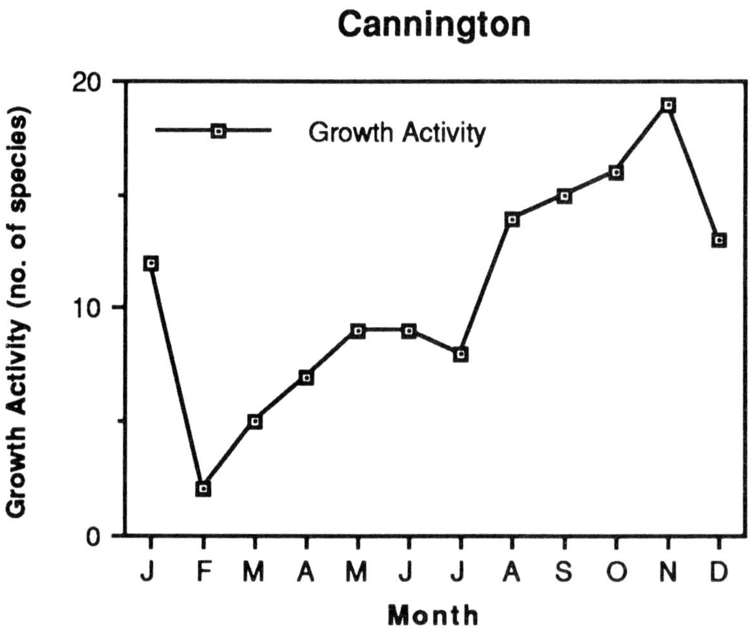

Fig. 3. Number of plant species (max. = 23) exhibiting growth activity in *Banksia* woodland at Cannington, near Perth (1978 data from Stephens 1979); data for the Kellerberrin area are unavailable.

area into two groups, winter-growing species from the lower-shrub or ground cover strata and spring-summer-growers. He hypothesized that the first group consists of shallow-rooted plants whose growth is governed by moisture availability, while the second group consists of deeper-rooted plants which have access to a more seasonally reliable water supply. Provided that water is available, growth is probably governed by adequate temperatures, solar radiation or an appropriate day length.

At both Karragullen (near Perth) and near Kellerberrin, flowering of plants is highly seasonal, with a peak in flowering around October (Fig. 4). No data are available on fruiting and seed production at Karragullen, but near Kellerberrin the main fruiting period took place in November, one month after flowering and maximum seed release was during November–February, 1–4 months after flowering (Fig. 4).

Litterfall values are available for both Dwellingup and for near Kellerberrin. The Dwellingup site is jarrah *(Eucalyptus*

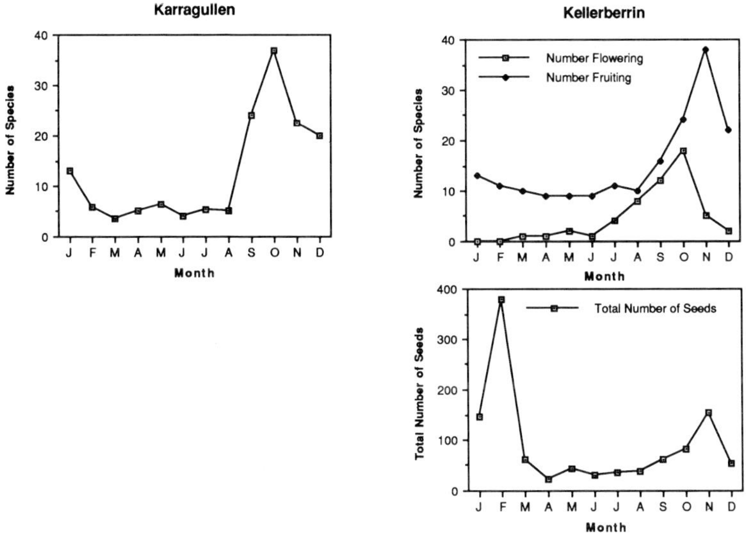

Fig. 4. Number of plant species observed flowering per monthly visit at Karragullen, near Perth (1978 data from Majer 1981) and near Kellerberrin (1986 data from van Schagen 1986). The number of species fruiting and the number of seeds collected in seed trays is also shown for the site near Kellerberrin.

marginata) open forest while the site near Kellerberrin is wandoo *(Eucalyptus capillosa)* woodland. At both sites litterfall takes place throughout the year, but peak fall occurs in summer when temperatures are high and strong easterly winds dislodge foliage (Fig. 5).

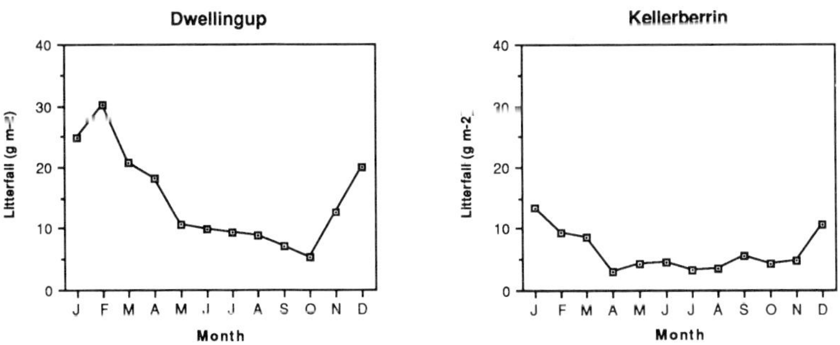

Fig. 5. Litterfall values for Dwellingup (1981 data from Postle *et al.* 1986) and near Kellerberrin (1991 data from H. C. Park, pers. comm.).

Litter-dwelling Collembola are predominantly members of the decomposer functional group. Data on Collembola phenology from Dwellingup indicate that this is closely tied to water availability (Fig. 6). There is a peak in mid-winter and activity virtually ceases during summer when the litter is dry. Corresponding data for the Kellerberrin area are not available, but incidental observations indicate that Collembola activity is more closely restricted to the winter months.

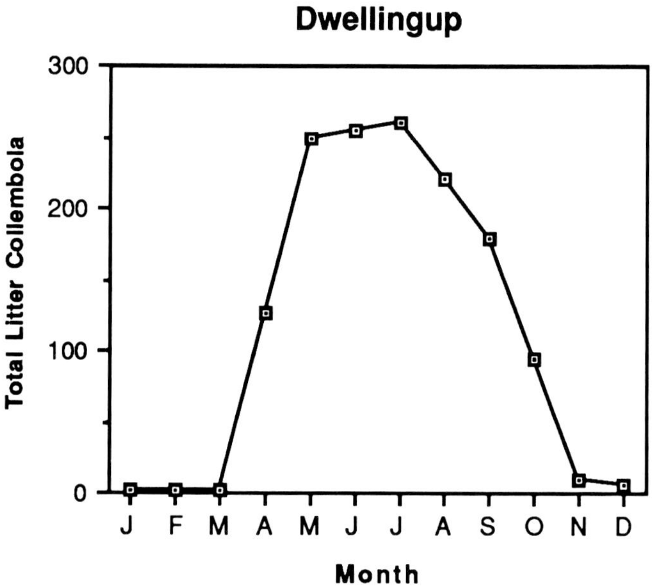

Fig. 6. Total litter-dwelling Collembola extracted from samples of leaf litter using a Berlese funnel (1981 data from Postle 1989); data for the Kellerberrin area are unavailable.

The majority of Hemiptera belong to what can broadly be described as the herbivore functional group. Sticky trap data for Serpentine, between Perth and Dwellingup, indicate a winter trough in Hemiptera biomass with a rise in spring through to a summer peak (Fig. 7). This pattern closely ties in with the major period of plant growth (Fig. 3). No comparable data are available for the Kellerberrin area.

The Diptera within the canopy of eucalypts comprise many individuals associated with pollination (e.g., members of the families Mycetophilidae, Ceratopogonidae and Syrphidae,

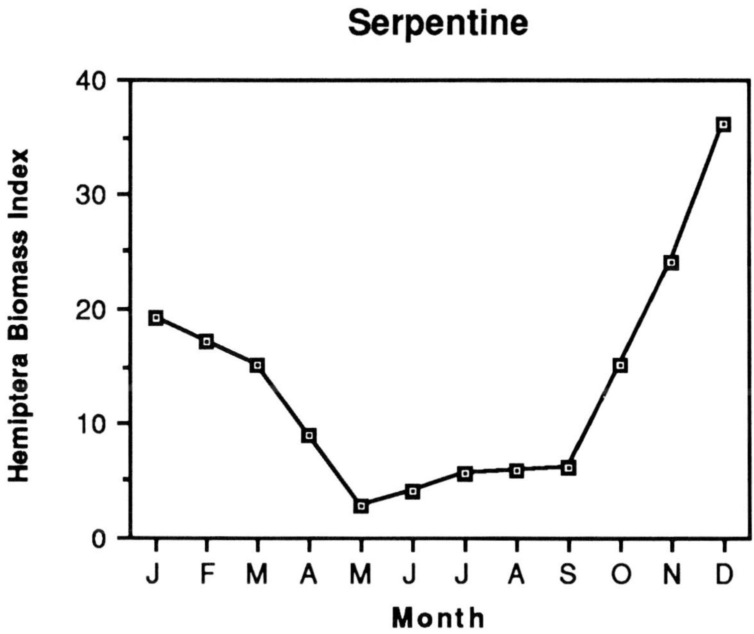

Fig. 7. Hemiptera biomass values obtained by sticky trapping at Serpentine, which is between Perth and Dwellingup (1981 data from Wykes 1985); data for the Kellerberrin area are unavailable.

(A. C. Postle and J. D. Majer, unpubl. data)). Consequently, a census of the overall abundance of Diptera might provide an index of pollinator activity. Figure 8 shows the variation in biomass of Diptera obtained from sticky traps set in a jarrah canopy at Serpentine, between Perth and Dwellingup. There is a distinct peak in activity between September and October which closely corresponds to the peak in flowering (Fig. 4). It therefore appears that the activity of pollinators closely reflects the phenological pattern of the plant community.

Ants cannot be ascribed to any one functional group, since they contain predators, scavengers, saprovores and seed-feeding species. Activity at both Dwellingup and Kellerberrin indicates an autumn–winter trough in numbers of species and individuals, with a spring–summer peak (Fig. 9). Although this pattern is partly tied to temperature, it is also related to food availability. The summer peak closely corresponds with seed production and may in part be accounted for by the availability of this component of the ant community's diet.

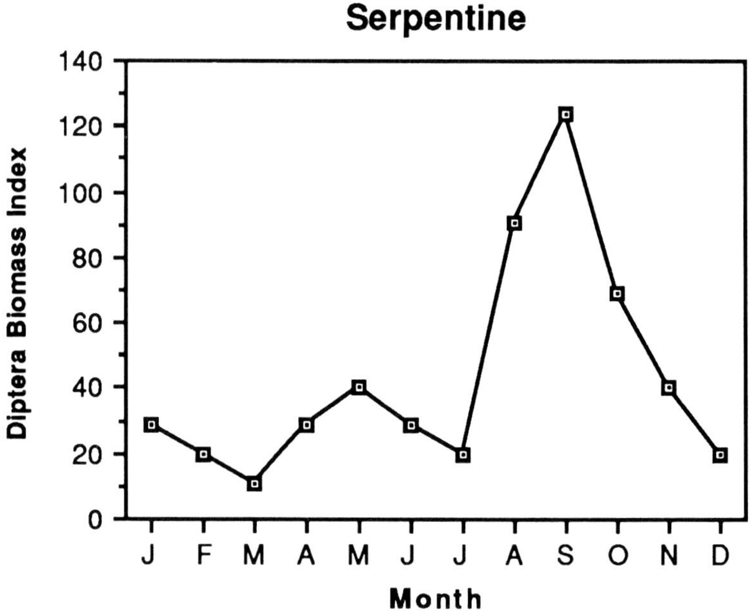

Fig. 8. Diptera biomass values obtained by sticky trapping at Serpentine, which is between Perth and Dwellingup (1981 data from Wykes 1985); data for the Kellerberrin area are unavailable.

Comparison of various seasonal patterns from a range of sites, collected during different years presents certain limitations. However, despite this, clear seasonal patterns in the climate, the plant community and the various animal functional groups can be seen. These are summarized for the "Perth"-site in Figure 10, which is simply a collection of the previous graphs. It has been arranged to show how the phenophases of the various functional groups are tied to the climate or the stage of activity of the plant community. It is evident that the phenology of the various functional groups is very clearly demarcated. Decomposition succeeds litterfall and is tied in with moisture availability, herbivory parallels foliage growth, pollination parallels flowering, and seed removal (as indicated by ant activity) corresponds to the main period of seed production. It has not been possible to provide a similar diagram for the Kellerberrin area because some of the datasets were not available. However, from the available data and my impressions derived from working in the area, it appears that the phenophases of the various functional groups are shorter and closely matched to the briefer periods of favourable climate or plant activity. It is this strong demarcation of phenological activity which I believe

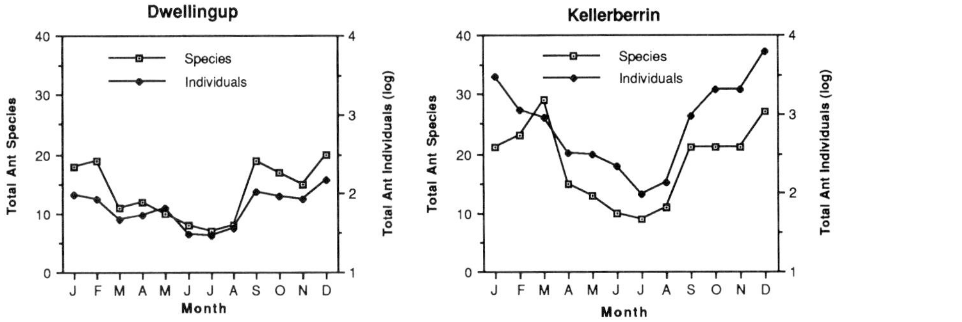

Fig. 9. Total ant species and individuals obtained by monthly pitfall trapping at Dwellingup (1976 data from Majer and Greenslade 1988) and near Kellerberrin (1986 data from van Schagen 1986).

influences the inertia and resilience of the mediterranean communities within Australia. This is further explored in the next section.

Inertia and resilience in mediterranean regions of Australia

How does the inertia and resilience of Australian mediterranean ecosystems compare with other mediterranean areas or with other climatic regions?

Inertia

In view of the fact that much of the mediterranean area of Australia has been altered by clearing, the inertia of the component ecosystems is not particularly relevant — most of the original components of the associated plant and animal community are inevitably altered by this land use.

I am not aware of any comparative studies which look at inertia in mediterranean parts of Australia in relation to other parts of the world. However, I believe that the combination of drought and high temperatures in summer, along with certain types of disturbance, presents particular problems in mediterranean Australia. Figure 11 is a diagrammatic representation of environmental conditions over a two-year period in mediterranean Australia. The combination of summer drought and high temperatures creates adverse conditions during mid to late-summer. These conditions are, to an extent, moderated by the buffering effect of the tree canopy, the ground cover and the leaf litter.

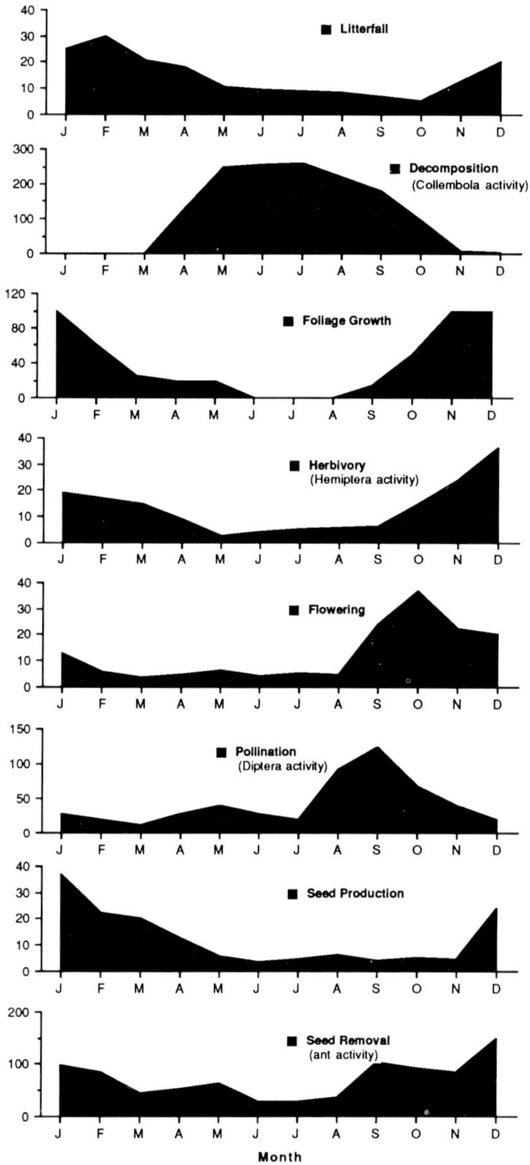

Fig. 10. Synthesis of the phenology data from the group of sites between Perth and Dwellingup, arranged to show how the phenophases of each functional group are tied to climatic or plant phenological factors and how the various phenophases of the various functional groups are not in phase with each other.

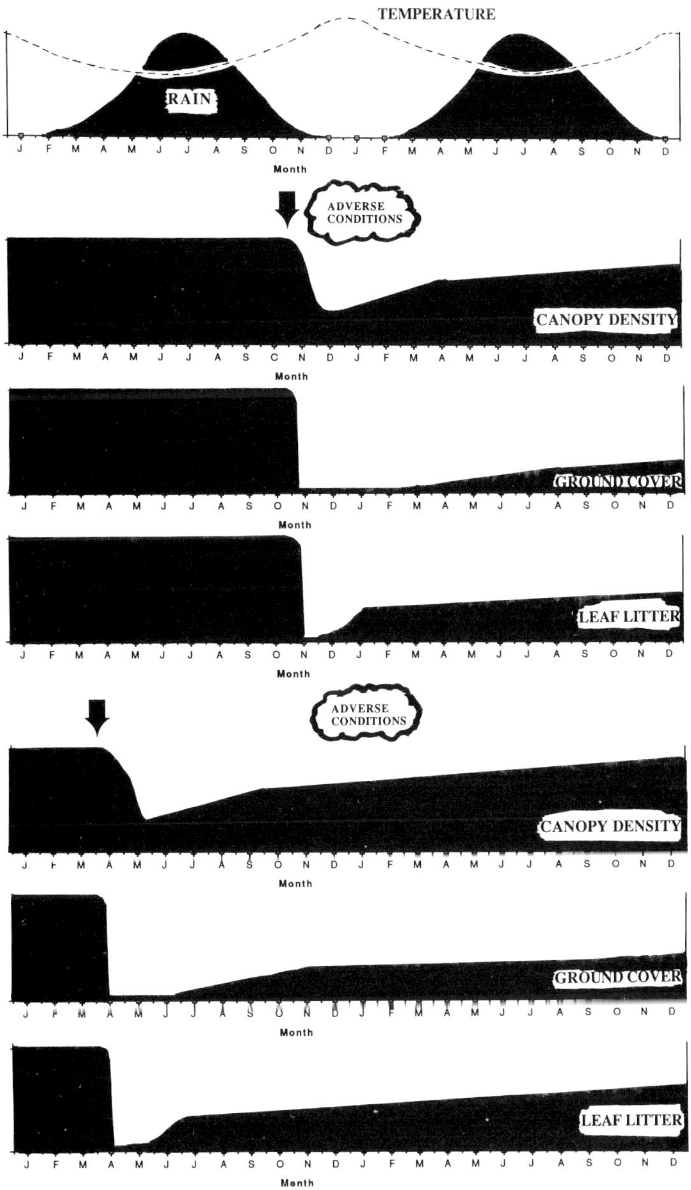

Fig. 11. Schematic diagram of climatic and plant cover values over a two-year period during which there is either an episodic spring or autumn disturbance (arrow).

These intercept some of the solar radiation, thus reducing temperatures and helping to maintain humid conditions.

Consider a disturbance during spring or early summer. This could be a fire, a logging activity, a tornado or some human-caused damage to the vegetation. To varying extents, these disturbances destroy the canopy, the ground cover and, in the case of fire, the litter layer. This lessens the ability of the ecosystem to buffer the microclimate at precisely the time of greatest adversity. It is only towards the end of the summer that the vegetation and litter has started to build up and hence buffer the effects of the harsh climate. The adverse summer conditions, combined with an episodic disturbance in spring or early summer, may, therefore, have a particularly severe impact on the biota.

By contrast, an episodic disturbance during autumn occurs at a time when the microclimate is cool and moist. This is followed by a period of regrowth of the canopy, the understorey and, in the case of a fire which causes crown-scorch, a build up of the litter layer. Thus, when the adverse summer conditions arrive, the ecosystem has once again developed some capacity to buffer the microclimate. It may, therefore, be expected that an autumn disturbance should have a lesser impact on the biota than one in spring. The data to substantiate this suggestion are lacking, despite the frequently cited controversy that prescribed burns in spring have a greater impact on the biota than those in autumn. Majer (unpubl. data) has found that the ant community was changed to a greater extent and took longer to recover after a spring burn than after an autumn burn. However, the spring burn was hotter than the autumn one, so this result could just as easily be explained in terms of fire intensity. On the other hand, Hobbs and Atkins (1990) found that in *Banksia* woodland, seedling regeneration occurred only in an autumn burnt area, while vegetative regrowth was more rapid in the spring burnt area. There are clearly differences in the response of the biota to spring and autumn disturbances (see also Lamont, Chapter 6) but the relative severity of disturbances during these two seasons requires more careful evaluation under controlled conditions and by investigating a whole range of components of the biota, representing the various functional groups.

Resilience

Fox and Fox (1986) have already compared the resilience of plant and animal communities between the various mediterranean regions of the world and also with other non-mediterranean ecosystems. They compared the results of various published, but not standardized, studies throughout the world. They concluded that:

1. Mediterranean regions are more resilient than most non-mediterranean regions in response to the human disturbances of fire and mining.
2. There is little evidence for differences in responses of the different mediterranean regions of the world to human disturbance.
3. There are differences in community response to different types of disturbance. In a given habitat, recovery after fire is more rapid than after clearing, which is in turn more rapid than after mining.

These authors completed their review by erecting a more specific hypothesis that "The stress produced by summer drought and frequent natural disturbance (rather than a mediterranean climate *per se*) has played an evolutionary role in producing communities that are more resilient to human disturbance." The documented studies reviewed by Fox and Fox (1986) do not yield conclusive trends and the authors conclude with the call for rigorous testing of this hypothesis.

Although a number of successional studies have since been carried out, I am still not aware of many studies which have been conducted in such a way that different mediterranean ecosystems may be directly compared with each other or with other regions of the world. My own work on ant recolonization of rehabilitated mines provides some data which are comparable between areas.

At various sites through Australia, ants have been surveyed in a chronosequence of rehabilitated areas and also in controls of native vegetation. Sites have also been surveyed in subtropical parts of Brazil and South Africa. In all studies, a standardized method of survey was adopted so the results may be compared. Comparison of species richness of ants in three-year-old rehabilitation throughout Australia reveals a gradient in ant return rate, from lowest in the drier mediterranean climatic zones to highest in the tropical monsoonal areas (Table 2) (Majer 1989). Furthermore, the return rates in Brazil and South Africa are in line with those for areas within Australia which experience a similar climate (Table 2). Species richness in the control sites within Australia and in the other two surveys were all within a fairly narrow range (about 30 species per survey-transect) so the results are not confounded by differences in α-diversity between the various sites.

Table 2. Estimated number of ant species found in three-year-old rehabilitated plots within mined areas situated throughout Australia, and in sites within Brazil and South Africa (adapted from Majer 1989, 1992; Majer and de Kock 1992).

Location	Mine type	Climate type	Annual rainfall (mm)	Number of ant species
Eneabba, WA	Sand	Mediterranean	550	6.0
Collie, WA	Coal	Mediterranean	973	8.0
Jarrahdale/Del Park, WA	Bauxite	Mediterranean	1 155–1 287	10.0
Myall Lakes, NSW	Sand	Warm-temperate	1 205–1 362	11.0
Poços de Caldas Brazil	Bauxite	Sub-tropical	1 695	14.5
Richards Bay, South Africa	Sand	Sub-tropical	1 292	18.0–22.0
North Stradbroke Island, Qld	Sand	Sub-tropical	1 645	18.0
Gove, NT Groote Eylandt, NT Weipa, Qld	Bauxite or Manganese	Tropical monsoonal	1 277–2 083	22.0

Since this trend in ant return rates shows up through a backdrop of widely differing rehabilitation techniques, it appears that climate has an overriding influence on the rate of ant return; in other words on the resilience of this component of the animal community. Although many aspects of the climate are involved (e.g., season of rainfall, length of rain period, etc.) the mean annual rainfall correlates well with ant species richness in three-year-old rehabilitation in Australia ($r = 0.92$, $p < 0.01$). The climate probably influences ant return directly, by mediating the microclimate in which the animals live, and indirectly, by influencing the plant variables with which the animals are coupled (Majer 1989). If this result is general, the potential climatic changes which are discussed by Saunders and Hobbs (Chapter 4) could have significant effects on ecosystem resilience.

These findings conflict with those of Fox and Fox (1986) in that resilience of the ant community is lowest in the mediterranean regions and it is even lower in the drier mediterranean regions. These observations apply to ants, although incidental observations made whilst carrying out the field work suggest that many other invertebrate groups are exhibiting similar trends. This, therefore, leads to the dilemma as to who is right? Is the biota in general less resilient in mediterranean regions than in certain other climatic zones (at least within Australia) or is this finding peculiar to ants, with the remainder of the biota being more resilient in mediterranean Australia than in other climates? The suggestion that biota in mediterranean regions is less resilient than in certain other regions is certainly consistent with the fact that much of the Australian biota evolved in more humid environments and that, since the Late Tertiary, these environments have experienced a drop in temperature and increasing aridity (Specht 1981a,b). The biota may, therefore, have retained features which are adapted to humid rather than arid environments. I believe that only after standardized surveys of community resilience are carried out in different climatic zones of Australia, preferably by the same groups of researchers, will it be possible to answer this question. Until then, we are left with unsubstantiated viewpoints and untested hypotheses concerning this question.

Maintenance and restoration of function

The ultimate question which is relevant to the theme of this book is whether diversity and ecological function can be maintained in the face of disturbance of the ecosystem and whether it can be restored once it has been reduced or lost.

There is no doubt that the rates of certain ecological functions are changed by disturbances such as clearing. For instance, soil structure is no longer maintained when land is cleared of its native vegetation and impacted by livestock and machinery. This is partly due to the loss of those large soil animals which normally maintain soil structure through their digging and burrowing activities (Abbott *et al.* 1979). Another example is that of herbivory, which is reduced for a time after burning because populations of animals such as grasshoppers are reduced and take time to reinvade (Whelan and Main 1979). No doubt there are instances where the species diversity of plant communities is reduced as a result of pollinator loss in disturbed areas. However, at present there are few experimental tests of this logical expectation in the literature (Whelan 1989). Finally, seed removal and dispersal is known to be initially reduced in areas which have been cleared, mined and subsequently rehabilitated (Majer 1980).

These examples, all taken from mediterranean parts of Western Australia, indicate that the rates of certain ecological functions can be reduced, at least for a period of time, following disturbance. There are, however, examples where function can be maintained or restored in areas which have respectively been disturbed or rehabilitated. To illustrate this I will give two examples from

Western Australia which concern the linked functions of litter decomposition and nutrient cycling.

The canopy and much of the understorey of jarrah forest is destroyed as a result of the pathogenic organism *Phytophthora cinnamomi*. In severe outbreak areas, as a result of the reduced plant cover, litter fall and litter mass are reduced and the micro-climate at ground level becomes more severe than in healthy forest. This causes a reduction in the densities (and no doubt diversity) of the soil and litter fauna. Jarrah leaves, which were confined in mesh bags and placed out on the ground for 21 months, lost more mass in the diseased than in the healthy forest. Since this was probably not accounted for by increased activity of the invertebrate fauna, it is postulated that the reduced abundance of decomposer-associated invertebrates was compensated for by the enhanced physical weathering of the litter (Postle *et al.* 1986). Here then is an example of one ecological function, decomposition, being maintained when the abundance and diversity of agents which are normally associated with this activity are reduced.

The second example relates to bauxite mining in Western Australia. Following clearing of the original jarrah forest, the ore is removed, topsoil is replaced and the area is rehabilitated with largely native vegetation. Greenslade and Majer (in press) have studied the recolonization of Collembola in a chronosequence of rehabilitated areas ranging from 0 to 13 years old. Sixty-nine species of Collembola were found in the rehabilitated areas but nine of the 28 species found in the forest control were absent in the rehabilitation. Development of the collembolan community was positively correlated with age of rehabilitation, plant species richness and percentage plant cover. Ward *et al.* (1991) investigated decomposition of jarrah litter in the same plots by using the same litter-bag technique that was described above. Litter decomposition in the better-developed rehabilitation was at a similar level to in the forest controls. Linear trends were found between litter decomposition on rehabilitated mines and understorey cover density, litter cover and soil moisture. The trends were similar to those of the Collembola and suggest that restoration of decomposition and this decomposer-associated component of the fauna are closely linked. It therefore appears that it is possible to restore this ecological function in areas which have been drastically disturbed. Further study is required to establish whether collembolan diversity affects decomposition rates and thus gives direct evidence of the effect of biodiversity on function. (See also Lambeck, Chapter 7.)

Conclusion

This review has concentrated on examples from mediterranean Australia, particularly Western Australia. It has provided some ideas on the vulnerability of the biota, and the associated ecological processes to disturbance. In addition, it has provided some comparative data on the resilience of certain components of the biota and one ecological function. However, it is quite evident that well planned, standardized studies which produce data suitable for comparison between areas are generally lacking. The only other data-sets from Australian-mediterranean environments which enable comparisons between plant and animal phenological events and subsequent discussion on the resilience of the community are those reported on by Specht *et al.* (1981) and Edmonds and Specht (1981). I believe that the debate on the resilience and inertia of mediterranean ecosystems and the role of biodiversity in these, especially in relation to those in other climatic zones, would be greatly enhanced if a series of standardized studies was initiated.

References

Abbott, I., Parker, C. A. and Sills, I. D., 1979. Changes in the abundance of large soil animals and physical properties of soil following cultivation. *Aust. J. Soil Res.* **17**: 343–53.

Beard, J. S. and Sprenger, B. S., 1984. Geographical data from the vegetation survey of Western Australia. Vegetation Survey of Western Australia. Occasional Paper No. 2. Vegmap Publications: Perth.

De Angelis, D. L. and Waterhouse, J. C., 1987. Equilibrium and non-equilibrium concepts in ecological models. *Ecol. Monogr.* **57**: 1–21.

Dell, B., Hopkins, A. J. M. and Lamont, B. B. (eds), 1986. Resilience in mediterranean-type ecosystems. Junk: Dordrecht.

Edmonds, S. J. and Specht, M. M., 1981. Dark Island heathland, South Australia: faunal rhythms. Pp. 15–27 *in* Ecosystems of the world: 9B Heathlands and related shrublands ed by R. L. Specht. Elsevier: Amsterdam.

Fox, B. J. and Fox, M. D., 1986. Resilience of animal and plant communities to human disturbance. Pp. 39–64 *in* Resilience in mediterranean-type ecosystems ed by B. Dell, A. J. M. Hopkins and B. B. Lamont. Junk: Dordrecht.

Greenslade, P. and Majer, J. D. (in press). Recolonization by Collembola of rehabilitated bauxite mines in Western Australia. *Aust. J. Ecol.*

Hobbs, R. J. and Atkins, L., 1990. Fire-related dynamics of a *Banksia* woodland in southwestern Western Australia. *Aust. J. Bot.* **38**: 97–110

Laut, P., Heyligers, P. C., Keig, G., Loffler, E., Margules, C. and Scott, R. M., 1977. Environments of South Australia handbook. CSIRO Division of Land Use Research: Canberra.

Majer, J. D., 1980., The influence of ants on broadcast and naturally spread seeds in rehabilitated bauxite mined areas. *Rec. Rev.* **3**: 3–9.

Majer, J. D., 1981. A flowering calendar for Karragullen, a northern jarrah forest locality. *WA Herb. Res. Notes* **5**: 19–28.

Majer, J. D., 1989. Long-term colonization of fauna in reclaimed land. Pp. 143–74 *in* Animals in primary succession: the role of fauna in reclaimed lands ed by J. D. Majer. Cambridge University Press: Cambridge.

Majer, J. D., 1992. Ant recolonization of rehabilitated bauxite mines of Poços de Caldas, Brasil. *J. Trop. Ecol.* 8: 97–108.

Majer, J. D. and de Kock, A. E., 1992. Recolonization of sand mines near Richards Bay, South Africa: an evaluation of progress with rehabilitation. *South African J. Sci.* **88**: 31–36.

Majer, J. D. and Greenslade, P., 1988. Soil and litter invertebrates. Pp. 197–226 *in* Mediterranean-type ecosystems: a data source book ed by R. L. Specht. Kluwer: Dordrecht.

Postle, A. C., 1989. Variations in soil and litter invertebrate populations within the northern jarrah forest. Ph.D. thesis, University of Western Australia.

Postle, A. C., Majer, J. D. and Bell, D. T., 1986. Soil and litter invertebrates and litter decomposition in jarrah *(Eucalyptus marginata)* forest affected by jarrah dieback fungus *(Phytophthora cinnamomi)*. *Pedobiologia* **29**: 47–69.

Saunders, D. A., Hopkins, A. J. M. and How, R. A. (eds), 1990. Australian ecosystems: 200 years of utilization, degradation and reconstruction. *Proc. Ecol. Soc. Aust.* **16.**

Specht, R. L., 1981a. Ecophysiological principles determining the biogeography of major vegetation formations in Australia. Pp. 301–33 *in* Ecological biogeography in Australia ed by A. Keast. Junk: The Hague.

Specht, R. L., 1981b. Evolution of the Australian flora: some generalizations. Pp. 785–805 *in* Ecological biogeography in Australia ed by A. Keast. Junk: The Hague.

Specht, R. L., Roe, E. M. and Boughton, V. H., 1974. Conservation of major plant communities in Australia and Papua New Guinea. *Aust. J. Bot. Suppl.* **5.**

Specht, R. L., Rogers, R. W. and Hopkins, A. J. M., 1981. Seasonal growth and flowering rhythms: Australian heathlands. Pp. 5–13 *in* Ecosystems of the world: 9B Heathlands and related shrublands ed by R. L. Specht. Elsevier: Amsterdam.

Stephens, L., 1979. Phenological responses of Banksia woodland species. M.Sc. thesis, University of Western Australia.

van Schagen, J., 1986. Ant-plant interactions in a Western Australian wheatbelt nature reserve. Graduate Diploma in Natural Resources thesis, Curtin University of Technology.

Ward, S. F., Majer, J. D. and O'Connell, A. M., 1991. Decomposition of eucalypt litter in rehabilitated mines. *Aust. J. Ecol.* **6**: 251–57.

Westman, W. E., 1986. Resilience: concepts and measures. Pp. 5–19 *in* Resilience in mediterranean-type ecosystems ed by B. Dell, A. J. M. Hopkins and B. B. Lamont. Junk: Dordrecht.

Whelan, R. J., 1989. The influence of fauna on plant species composition. Pp. 107–42 *in* Animals in primary succession: the role of fauna in reclaimed lands ed by J. D. Majer. Cambridge University Press: Cambridge.

Whelan, R. J. and Main, A. R., 1979. Insect grazing and post-fire plant succession in south-west Australian woodland. *Aust. J. Ecol.* **4**: 387–98.

Wykes, B. J., 1985. The jarrah avifauna and its re-establishment after bauxite mining. *W.A.I.T. School of Biology Bulletin* **11**.

PLATE 6

LEFT: Does diversity have an ecological function? What role do these spider orchids have in mediterranean ecosystems? (Photo: R. J. Hobbs).

BELOW: The galah *Cacatua roseicapilla* has invaded mediterranean regions from the arid interior and has a significant impact on other bird species and the vegetation by ringbarking mature trees and grazing on seedlings. (Photo: R. J. Hobbs).

CHAPTER 10

Management to retain biodiversity in the face of uncertainty

A. R. Main[1]

This chapter discusses management in terms of decision making in the face of uncertainty and argues that formal decision theory and processes should be followed.

Introduction

Accepting that the retention of biodiversity is the goal of management, then the central problem becomes how to proceed with decision making to achieve the goal. The traditional way is to seek precise predictions about the future based on a detailed understanding of the nature of the problem, i.e., to be as near to certainty as possible before proceeding. Generally this is interpreted as requiring more research or monitoring before engaging in predictions or making management decisions. As will become apparent below this is an ideal which cannot be satisfied because of uncertainty about the future and another solution needs to be found. For comparisons of conventional and adaptive approaches see Walters (1986).

Large-scale environmental changes

Biodiversity in any given region will, in part, be dependent on the kind of climate experienced but in this regard the future is very uncertain. Increasing amounts of greenhouse gases in the atmosphere have the potential to produce marked changes in climate, but these cannot be modelled adequately on a regional scale. Currently two scenarios have been proposed for southern Western Australia (Hille 1989; see also Saunders and Hobbs,

[1]Zoology Department, University of Western Australia, Nedlands, Western Australia, Australia 6009.

Pages 193–209 *in* BIODIVERSITY OF MEDITERRANEAN ECOSYSTEMS IN AUSTRALIA ed by Richard J. Hobbs. Surrey Beatty & Sons, 1992.

Chapter 4). The first predicts an intensification of the summer monsoonal rainfall and a poleward shift in the westerlies. The second scenario also assumes a poleward shift in the westerlies but it also assumes an increase in the number of "cut-off" lows produced as a result of a more split flow pattern in the upper winds over the Australian region in winter (Hille 1989). For the south-west of Western Australia the first scenario suggests an increase of +1.2–1.5°C in summer and +1.8–2.1°C in winter, and scenario two suggests a +3.2–4.0°C increase in summer and a +4.8–5.6°C increase in winter. While the second scenario is at present considered less likely it does represent what could happen with a high climatic sensitivity or a large increase in greenhouse gas emissions as could happen if international control is not imposed or is ineffective (Hille 1989). Regional rainfall patterns are the most uncertain of all climatic variables but the first of the models used above suggests a decrease of up to 20 per cent in winter rainfall in the lower western part of the south-west. This reduction declines to the interior and north-east. The second scenario suggests a slight increase in the winter rainfall of the lower south-west. Both scenarios suggest an increase of up to 40 per cent in summer rainfall in the lower western interior.

The "El Niño" southern oscillation (ENSO) affects rainfall in Australia (Henderson-Sellers and Blong 1989; Broadbridge 1989; Evans 1990), and is associated with drought. It thus contributes to interannual variability; but whether, and in what sense, ENSO will be affected by a greenhouse warmed world is not clear (see Pittock 1988; Evans 1990). Thus we have no idea of whether variability will increase, decrease, or remain the same (Henderson-Sellers 1991). However, the biota is sensitive to episodic events rather than to averages and so is most responsive to seasons which are moister or drier than average, and it is the significance of this phenomenon that is discussed later with respect to successful maintenance of diversity.

The effect of settlement and exotics on diversity

The immediate effect of settlement has been the catastrophic decline in area occupied by, and abundance of, much of the biota. The long-term effect on the biota of fragmentation of the natural environment following settlement and clearing is uncertain but may well result in reductions in range and abundance of many species (see Saunders and Hobbs, Chapter 4). The need for new techniques to control the exotics which accompanied settlement and have seriously reduced the diversity in the natural biota (see

Main, Chapter 5) is acknowledged; but their availability in the near future is unlikely. All these are among the uncertain matters that must be considered when reaching decisions regarding management procedures for retaining biodiversity. In the present case the matter comes down to what should be researched to reduce uncertainties before making management decisions. A call for more research without knowing what deficiencies in knowledge are most important (in the context of current decision making) is likely to lead either to a continuation of established workers' research or the beginning of work that has intrinsic intellectual appeal only.

It has been argued (Main, Chapter 5) that retaining biodiversity has a significance beyond the moral, cultural, aesthetic or "environmental-ethical" considerations justifying retention of other organisms (see Hobbs, Chapter 1) by explicitly suggesting that the retention of biodiversity is the only insurance we have that ecosystems will continue to function more or less in their present form as climate and environmental conditions change in the future.

Political and social boundaries

The Western Australian Government has endorsed the national conservation strategy for Australia (Anon. 1984), is committed to the objectives of the national strategy and has prepared a state conservation strategy (Anon. 1987). Among the goals of that strategy are: "to maintain essential ecological processes and life support systems and to preserve genetic diversity" (Anon. 1987), which is, of course, reflected in the variety of genotypes in individuals, populations and species.

In the wheatbelt of Western Australia the problem of maintaining biodiversity is embedded in the social and economic problems of an agricultural society under threat from encroaching salinization, wind and water erosion. These are interconnected and are the result of destruction of native vegetation which, before settlement, impeded the overland flow of wind and water and provided a water pump which maintained the hydrological balance of the soil. In this broader context restoring biodiversity by replanting trees contributes to the economic viability of the region by reducing the loss of agricultural land through erosion and salinization. Maintaining, or re-establishing, biodiversity is thus part of the solution to the problem of sustainable development.

There thus happens to be a timely congruence between the ecological justification for retaining biodiversity (Chapter 5) and a political and social recognition that living resource conservation is not only a culturally acceptable goal but also provides services to the farming community (see Fry and Main 1993; Lefroy *et al.* 1993). These considerations set the boundaries of the problem. They extend beyond what would normally be considered the limits of a strictly biological problem to embrace political, social and biological justification for making management decisions which will ensure the retention of diversity in the future, while we lack, and are never likely to have, full knowledge of the present and are faced with massive uncertainty regarding the future.

The decision making process

In the above circumstances it is unreasonable to expect "correct" management decisions but it is possible to make rational decisions in the face of uncertainty. Most management decisions based on predictions from models proceed as though uncertainty had been resolved in the prediction process. Modelling can clarify and highlight these uncertainties but cannot resolve them, and ultimately many management decisions are essentially gambles (Walters 1986). The alternative is to embrace uncertainty, construct alternative models and place odds on them (Campbell 1967; Raiffa 1968). This sort of model making has two functions: first, to expose uncertainties and stimulate thinking about alternative options; and second, to rank and compare options identified (Walters 1986).

The way individual judgments about uncertainties may be encoded as probabilities is discussed by Spetzler and Staël Von Holstein (1975) in a business context, but the methodology has a wide application. These personal subjective assessments, based on beliefs and preferences, are referred to as conditional probabilities (Campbell 1967; Raiffa 1968) or prior probabilities (Walters 1986) and express the uncertainties about the final outcome. They are, in effect, personal statements that some outcomes seem more credible than others. The probabilities derived in this way are then manipulated using Bayes's theorem (Campbell 1967; Raiffa 1968; Walters 1986; Howson and Urbach 1991).

Decision theory recognizes that decisions have two components; the choice of a way to proceed and the effect of chance in determining the outcome. Decision trees or flow diagrams are developed in order to present the alternatives as simple, easy to visualize, models. They are arranged in such a way as to represent, in chronological order, on each pathway representing an alternative course of action, a decision made followed by the probabililty that the desired outcome will arise (Raiffa 1968). Such rational procedures have two consequences: inadequate knowledge is manifest and decisions can be modified as soon as new or more precise information is gained.

Decision theory has many potential applications (see Behn and Vaupel 1982; Bell *et al.* 1988; Manne and Richels 1991) but has not been used much in biological problems, although its use was illustrated in the biologically oriented statistics text by Campbell (1967). This is perhaps because ecologists have been concerned more with describing nature than prescribing actions to manage it (but see Maguire 1988, 1991; Maguire *et al.* 1983; Walters, 1986; Valentine *et al.* 1976).

Assessing the success of management to retain biodiversity is possible only when there is an objective way of measuring it. As a term biodiversity is too diffuse in its usage and so is not helpful for assessing whether management has retained it or not. Thus it becomes necessary to select some way of measuring succesful retention of biodiversity. Suppose that, after discussion within the group concerned with selecting management options, it was accepted that an assumption that the number plant species present would mirror the diversity of other organisms and so be an objective measure. Then a count of the number of species present now would provide a baseline and scoring changes in their numbers following management would be an objective measure of the success or otherwise of management.

A formulation in terms of decision theory would consider things which could affect the number of species present as arising from management decisions or chance events. These are broadly of two kinds; mortality due to other organisms such as grazing and predation; or meteorological conditions such as weather or climate. This formulation of the problem assumes that reduction in species diversity arises from local extinctions which follow from long-term failure of recruitment. A species count may not be absolute because in some cases species may become apparently extinct while still persisting within a soil seed bank. Such cases are not a high proportion of the local flora and their presence is unlikely to impair the use of counts of plant species as an objective measure.

Climate and meteorological conditions affect mortality by reducing the life expectancy of individuals through early death, or effective recruitment by reducing the success of regeneration or establishment of juveniles. Such effects can arise from intense extreme events such as drought, very high temperatures, or strong hot winds. These may be associated or occur independently. The return time of such events is important, particularly in relation to generation time and life histories of species. Should the return time be frequent it will have an effect on the persistence of species whose life cycle traits are incompatible with the recurrent event-induced mortality, and so it will ultimately affect the composition of the biota. Thus when devising management plans it is necessary to consider events that are likely to affect the retention of species and assign prior probabilities to their effects, to devise management actions that can be taken, and to assign prior probabilities to take account of the likely effect of chance in determining their likely success.

The group charged with devising management procedures which will retain biodiversity needs to bound the problem socially and ecologically, consider the chance of environmental causes of failure, and characterize success in an objective way. From this, it will be apparent that the problem is so complex and has so many ramifications that it must be condensed or simplified so that the consequences of alternative decisions and chance events can be modelled and hence visualized, a preferred outcome can be chosen and management action can be selected.

Simplification

In earlier sections it has been argued that changes in mortality and recruitment are likely to affect the composition of the flora. Assuming that this is so allows for a first simplification, namely that likely increases in rainfall (Hille 1989) in southern Western Australia in a greenhouse world can be ignored as a factor likely to cause loss of species. The two scenarios proposed by Hille (1989) suggest that higher average temperatures are likely. He further suggested that the conditions of the summer of 1988, when the sequence of high temperatures reached an all time record, might become the norm with extreme temperatures above that. A similar extreme high temperature event occurred in January and February 1991 (Anon. 1991). On this occasion established and regenerating vegetation was killed by the strong, very hot winds associated with a heat trough along the west coast. Widespread fires are often associated with these conditions. Thus while increased rainfall may initiate germination and early establishment, extreme high temperatures may, on balance, be more significant. Hille's first scenario is thought to be more likely than the second (Hille 1989), and it would be reasonable to assign a higher prior probability to it than his second, warmer, scenario when higher mortality, and hence greater likelihood of extinction might be expected.

The social and economic consequences of loss of biodiversity are being approached by local farmers on the scale of water catchments through co-operative actions in the local Landcare Groups. These groups are attempting to control erosion and restore the hydrological balance by planting natural vegetation, perennial fodder crops, or trees to be harvested for timber. These activities mean that diversity is being restored to the agricultural landscape in the expectation that it will lead to the restoration of the hydrological balance and ultimately sustainable use of the land (see Hobbs, Chapter 11). Should it be possible to demonstrate that native vegetation is the equal of or more efficient than exotics in the long term, then a major gain of biodiversity will have been achieved. Thus simplified it is possible to see a unity in the objectives of economically and ecologically driven desires to retain biodiversity.

Biological assumptions

Normally regeneration takes place when light and space are made available, usually upon the death of the previous occupant of space. Currently increasing the availability of space is almost universally equated with burning, an opinion justified by arguing that the fauna and flora have adapted to regular burning over thousands of years of Aboriginal occupancy. However, release of space through canopy loss can arise from many causes such as death from old age, disease, wind throw (such as caused by tornadoes) and wild fires arising from lightning strikes. Currently loss from burning is commoner than all other causes of canopy loss because of two factors: the practice of hazard reduction burns which is assumed to imitate Aboriginal burning practice; and fires originating from acts of arson, which are often lit in hot, windy, dry and extremely hazardous conditions and result in very destructive burns. The foregoing suggests that fire-initiated regeneration may be common or usual in the future and is likely to lead to selection of biota which can tolerate regular burning. Possible losses may be anticipated if it is possible to observe and classify regeneration into that which occurs under canopy or in the open, and under or near, *versus* away, from the parent. Should any species require the presence of a canopy or parent for regeneration then such species will be disadvantaged by regular, especially non-patchy, burns. The significance of such a factor possibly contributing to the loss of biodiversity needs study. However, for the present purpose it is adequate to consider the factor which initiates regeneration as being canopy removal.

Good seasons following canopy removal may lead to good germination. However, this does not necessarily lead inevitably to re-establishment because of the effects of either extreme events such as drought or heatwave conditions as occurred in the summer of 1991 (Anon. 1991), or grazing by herbivores. The probabilities of climatic events may be derived from the two scenarios of Hille (1989). The scenario with moderate increases in temperature is thought to be most probable. However, Hille (1989) suggested that the extreme high temperature exerienced in the 1988 summer may become the norm, and summer drought and heat may cause deaths as occurred in 1991. In the case of the warmer, second scenario, even more extreme summer temperatures might be expected.

The effect of grazing in reducing the survival of regenerating plants will be related to the numbers of herbivores likely to be present. Experimental evidence is needed from replicated exclosures designed to distinguish the effects of grazing by macropods, rabbits, and grasshoppers. I assume that the effects will be as significant as those demonstrated on Rottnest Island in the 1950's when exclosures revealed that, after burning, grazing by the quokka *(Setonix brachyurus)* prevented the establishment of all but the most unpalatable plants (pers. obs.). Similarly, Whelan and Main (1979) have shown that, after burning, grass-hoppers can graze seedlings and suppress the regeneration of trees.

Biodiversity is related to the timing and initiation of regeneration of patches. The mosaic of different patches present in an area may be due to history of regeneration or soil type. In the past, episodes of changed climate resulted in these patches, and hence the biota, being shuffled over the countryside. In the present fragmented landscape this is no longer possible, and this increases the possibility of losses from the biota, a circumstance that emphasizes the need for an adaptive rather than a rule based approach to management.

Management

Main (1987) addressed the matter of managing and monitoring remnants by tabulating likely outcomes of alternative actions. The goal of managing to retain biodiversity embraces this but is more inclusive. Managers charged with maintaining biodiversity face choices between alternative courses of action when the necessary scientific information on which to make a choice is inadequate. Current biological research and research findings are not necessarily directed towards, or relevant to, answering problems faced by management. These inadequacies immediately suggest management oriented research topics, but again choices will have to be made because resources for research will always be inadequate. The problem of chosing between alternatives is central to management.

Finally managers, unlike research scientists, who can change theories when the current one is shown to be inadequate, are faced with the prospect that an inappropriate action may lead to loss of what they are charged to retain. Managers are thus cautious and prone to actions which display risk aversion as a prime consideration.

Possible decisions

Taking all the foregoing into account and recognizing the uncertainties involved the following appear to be some of the possible alternatives to be considered by managers when planning to retain natural diversity:

(a) Do nothing.

(b) More research, on what? Regeneration? Short-lived or long-lived plants?

(c) Monitor indicator species using transects? Quadrats? Replicated? Stratified?

(d) Work to conserve the natural environment independently of the economically driven endeavours of the farmers.

(e) Work co-operatively with the Landcare and other farm oriented groups.

(f) Seek other alternatives, e.g., set up standard monitoring sites of fenced exclosures by which the effects of grazing by herbivores can be measured and so determine whether, or what, vegetation will re-establish if grazing is reduced or absent.

Again design and replication are critical components of this alternative.

The first approach requires no management action and in the absence of an initial census will never be capable of determining what, if anything, has been lost. The second approach defers decisons about management but poses many in relation to what is to be researched. The monitoring has similar constraints; it will document the changes but defer management decisions to the future. The fourth offers more freedom of action but is handicapped because the opportunity to complement each approach, e.g., by enhancing wind erosion control, is lost. The fifth approach requires that time be devoted to joint understanding of goals and involves genuine co-operation but is likely to lead to achieving the maximum benefit to biodiversity. The last approach suffers from the same defects as b and c above and as with them the question arises as to whether the results will be timely, i.e., available before biodiversity is lost.

The uncertainties can be reduced to:

1. Climatic change; the likelihood of one or the other of the scenarios of Hille (1989) eventuating.
2. The likely effects of climatic change on the persistence and regenerative capacity of the biota. See Table 1 for the likely effects.
3. The likely effects that exotic predators and herbivores may have on diversity.
4. Actions which may reduce or ameliorate the effects in 1–3.
5. Probabilities of successful retention of biodiversity as a result of intervention.
6. Costs of ameliorating activities.
7. Preferences with respect to outcomes.

Summarizing from earlier sections and Table 1 the following possibilities can now be recognized:

(a) Climate 1.0–1.5 degrees warmer in summer suggesting some high summer temperatures (scenario 1).
(b) Climate 3.2–4.0 degrees warmer in summer suggesting frequent troughs along the west coast accompanied by extremely high temperatures and hot, strong, north-easterly winds (scenario 2).
(c) Overall the changes likely in (a) above are unlikely to affect seriously recruitment and the maintenance of biodiversity (Table 1).

Table 1. Management implications of possible effects of climatic change, following the two scenarios presented by Hille (1989). See text for details.

	Scenario 1	Scenario 2
Canopy loss	Fires local.	Fires intense, extensive, heat troughs, high temperatures.
Germination and regeneration	More or less as at present Regeneration good.	Less than at present. Regeneration poor.
Establishment	Drought and high temperatures have some effect on establishment.	Extremes of high temperatures more frequent, establishment poor.
Grazing	Myxomatosis widespread following cyclonic rains (Calaby *et al.* 1960); grazing less intense.	Myxomatosis only local following thunderstorms; grazing intense.
Effect on biodiversity	Reduced.	Much reduced.

(d) The changes likely under (b) above are very likely to impair recruitment and reduce biodiversity, analogous to moving towards the arid interior.

(e) Grazing by herbivores will reduce the chances of plant regeneration while predation by foxes and cats will eliminate many animals; together these will enhance the deleterious effects of either of the climatic changes and result in a massive reduction in biodiversity.

(f) Elimination, regulation or control of grazing is unlikely to be completely successful. Judgments of the likely success of such activities can be incorporated into decisions by expressing them as prior probabilities.

(g) Co-operating with, and inducing other organizations to revegetate with local species will materially enhance the prospect of retaining biodiversity.

Overall, and over the time span of decades necessary for regeneration, the likelihood of regeneration and hence maintenance of diversity under the assumptions of the biological effects made is higher for scenario 1 than for scenario 2. However, it should be remembered that the impact of exotics, especially herbivores, is unmeasured and the effect of predators only inferred from the extinctions of native mammals and the recovery of rock-wallaby populations after poisoning of foxes (Kinnear *et. al.* 1988).

Rational decisions cannot be made without considering the foregoing alternatives in an orderly manner. Normally this would be done by a workshop composed of people knowledgeable of the various aspects which may influence the outcome of the various alternatives. The difficulties at this stage are discussed by Walters (1986). The difficulties centre around the order in which decisions are to be made, the likelihood of chance determining the outcome and a judgment of the prior probability that chance will produce results one way or the other. The prior probabilities are measures of the credibility of an outcome; for example, it is not credible to assume that repeated high temperature events and fires as experienced during the summer of 1991 will have no effect on the biota. As a first approach the alternative courses of action and the sequences of choices (decisions) and chance can be set in a diagram or decision tree (Raiffa 1968). As an example and by way of illustration a partial tree (one of the alternatives which might be considered) is presented in Figure 1. In the example the convention is followed that decisions are represented by squares and chance events by circles.

At this stage it is possible to consider the preferences for the outcomes listed in Figure 1. Suppose it is considered that retention at no cost, c and g, were ranked highest but c was preferred to g because biodiversity may be lower in the latter case (the conditions assumed for scenario 2 are likely to lead to loss of species, see b and d above and Table 1). The other outcomes listed in Figure 1 may be ranked according to preference using the following rationale; retention at some cost be ranked above losses from ineffective grazing control and above losses due to no control. Then an order of preference might be c >g > a > b > e > f > d > h (when > is preferred to). Other rankings of the outcomes are of course possible depending on the values or preferences given to them. However, once values or preferences have been assigned it is assumed that they will not be altered during the modelling. Once preferences have been established it will become apparent that some outcomes are much preferred to others. These preferred outcomes can be emphasized by weighting them rather than having a simple ranking. These weighted values can be thought of as judgments of the "usefulness" or importance of the outcome in terms of the goal of retaining species, and may be termed the "utility" of the weighted, ranked, outcomes. Suppose that the weightings attributed to the outcomes were such that c had a utility of 12 with utility values declining so that h was assigned a utility of 1 (Table 2).

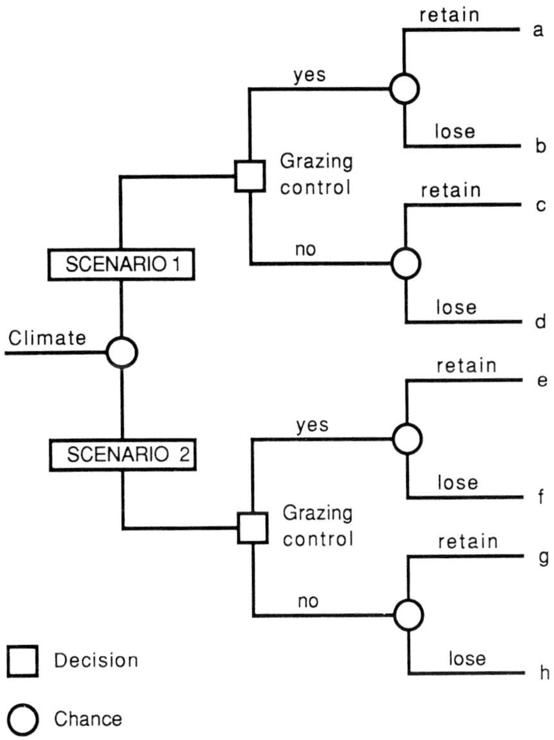

Fig. 1. Effects of two scenarios for climatic change, alternative courses of management action, and likely outcomes for retention or loss of species following these actions.

In the event that an outcome is not preferred then the difference between the utility assigned to it and the utility of that preferred can be said to be the "expected regrets" arising from the less than preferred outcome. Thus, c, the preferred outcome, has a utility of 12 and if it were to eventuate there would be no regrets. Utilities assigned to other outcomes are set out in Table 2.

Let it be assumed on the basis of the earlier discussion that prior probabilities of extinctions as a result of outcomes of climatic change, decisions taken, and chance are as set out in Figure 2. Then the prior probabilities and regrets can be combined to give a probable regret for each action under the uncertainties assumed.

Table 2. Outcome of alternative actions (from Fig. 1) listed in order of preferences with rationale for choice and the utilities and regrets associated with each choice (see text for details).

Outcome	Rationale	Utility	Regrets
c	No cost	12	0
g	As for c, but losses to climatic change	11	1
a	Retain at a cost	8	4
b	As for a, losses to ineffective grazing control	6	6
e	Retain at a cost, losses to climatic change	4	8
f	As for e, losses to ineffective grazing control	3	9
d	Losses from grazing	2	10
h	As for d, and to climatic change	1	11

These are presented as a decision tree in Figure 2 and suggest that the least probable regrets will occur when climate change has little effect on biodiversity and grazing control is implemented. This is a trivial result but serves to illustrate the method.

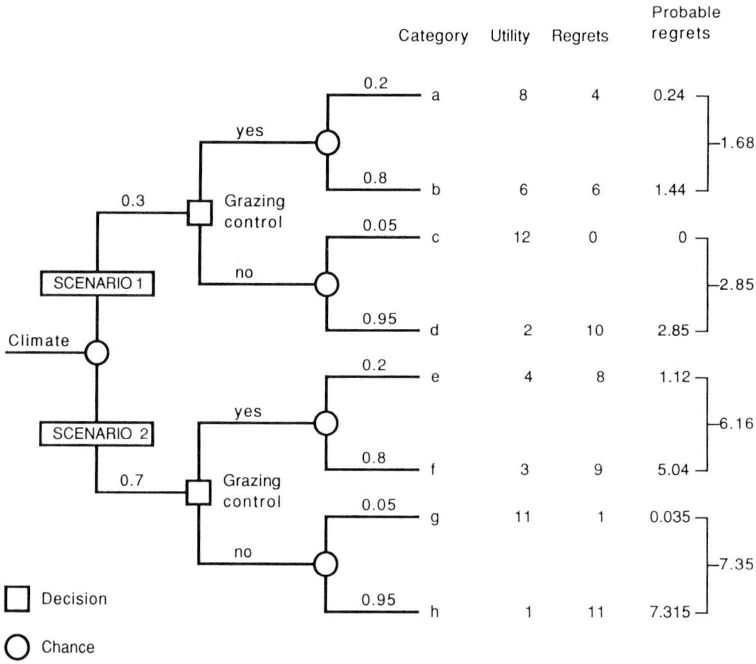

Fig. 2. Assumed prior probabilities that climatic change and choice of management action shown in Figure 1 will lead to the extinction of species, the utilities and regrets and probable regrets (see text) associated with each of those outcomes.

Discussion

The thrust of this chapter has been on a method of choosing between alternative management options. An effective and rational process is imperative when the risk of loss of biodiversity is high, and time and resources are limited. The process illustrated is effective and efficient in the sense that it ensures that not only are alternatives considered along with possible outcomes of pursuing that choice, but it also specifically admits that the future is uncertain. Rationality is achieved when choices between courses of action are made openly, as in workshops, which involve both managers and research workers. In the context of making choices between alternatives in a climate of uncertainty, formal decision processes, whether they be referred to as decision analysis (Raiffa 1968; Maguire *et. al.* 1983), decision models (Valentine *et al.* 1976), decision making (Goodall 1978), risk analysis (Maguire 1991) or adaptive management (Walters 1986) have many advantages, especially in making the assumptions and reasoning explicit. The urgency of the problem of the impending loss of biodiversity from mediterranean ecosystems and the high degree of uncertainty associated with the future suggests that management should actively engage in decision analysis as a means of choosing courses of action.

References

Anon., 1984. A national conservation strategy for Australia. AGPS: Canberra.

Anon., 1987. A state conservation strategy for Western Australia: a sense of direction. Bulletin 270, Department of Conservation and Environment: Perth.

Anon., 1991. Monthly weather review: Western Australia, February 1991. Bureau of Meteorology: Canberra.

Behn, R. D. and Vaupel, J. W., 1982. Quick analysis for busy decision makers. Basic Books, Inc.: New York.

Bell, D. E., Raiffa, H., Tversky, A., 1989. Decision making: descriptive, normative, and prescriptive interactions. Cambridge University Press: Cambridge.

Broadbridge, L., 1989. Climate change. Pp. 29–33 *in* Greenhouse 88: planning for climatic change ed by G. Lowe. Western Australian Water Resources Council and Environmental Protection Authority: Perth.

Calaby, J. H., Gooding, C. D. and Tomlinson, A. R., 1960. Myxomatosis in Western Australia. *CSIRO Wildlife Res.* **5**: 89–101.

Campbell, R. C., 1967. Statistics for biologists. Cambridge University Press: Cambridge.

Evans, J. L., 1990. Envisaged impacts of enhanced greenhouse warming on tropical cyclones in the Australian region. CSIRO Division of Atmospheric Research, Technical paper no. 20.

Fry, G. and Main, A. R., 1993. Restoring seemingly natural environments on agricultural land. *In* Nature Conservation 3: reconstruction of fragmented ecosystems, global and regional perspectives ed by D. A. Saunders, R. J. Hobbs and P. R. Ehrlich. Surrey Beatty & Sons: Chipping Norton, NSW. In press.

Goodall, D. W., 1978. Grazing management in Rajasthan: decision-making in the face of uncertainty. Pp. 81–94 *in* Glimpses of ecology ed by J. S. Singh and B. Gopal. International Scientific Publications: Jaipur, India.

Hendersen-Sellers, A., 1991. Greenhouse impacts and Australia. *Search* **22:** 210–13.

Henderson-Sellers, A. and Blong, R., 1989. The greenhouse effect: living in a warmer Australia. New South Wales University Press: Kensington, NSW.

Hille, R. J., 1989. Regional scenarios for climate change in WA. Pp. 33–35 *in* Greenhouse 88: planning for climatic change ed by G. Lowe. Western Australian Water Resources Council and Environmental Protection Authority: Perth.

Howson, C. and Urbach, P., 1991. Bayesian reasoning in science. *Nature* **350:** 371–74.

Kinnear, J. E., Onus, M. L. and Bromilow, R. N., 1988. Fox control and rock-wallaby population dynamics. *Aust. Wildl. Res.* **15:** 435–50.

Lefroy, E. C., Hobbs, R. J. and Scheltema, M., 1993. Reconciling agriculture and nature conservation: toward a restoration strategy for the Western Australian Wheatbelt. *In* Nature Conservation 3: reconstruction of fragmented ecosystems, global and regional perspectives ed by D. A. Saunders, R. J . Hobbs and P. R. Ehrlich. Surrey Beatty & Sons: Chipping Norton, NSW. In press.

Maguire, L. A., 1988. Decision analysis: an integrated aproach to ecosystem exploitation and rehabilitation decisions. Pp. 105–22 *in* Rehabilitating damaged ecosystems, Volume 2 ed by J. Cairns, Jr. CRC Press, Inc.: Boca Raton, Florida.

Maguire, L. A., 1991. Risk analysis for conservation biologists. *Conserv. Biol.* **5:** 123–25.

Maguire, L. A., Seal, U. S., Brussard, P. F., 1983. Managing critically endangered species: the Sumatran rhino as a case study. Pp.143–58 *in* Viable populations for conservation ed by M. E. Soulé. Cambridge University Press: Cambridge.

Main, A. R., 1987. Management of remnants of native vegetation. Pp. 1–13 *in* Nature Conservation: the role of remnants of native vegetation ed by D. A. Saunders, G. W. Arnold, A. A. Burbidge and A. J. M. Hopkins. Surrey Beatty & Sons: Chipping Norton, NSW.

Manne, A. A. and Richels, R. G., 1991. Buying greenhouse insurance. *Energy Policy* **19:** 543–52.

Pittock, A. B., 1988. Actual and anticipated changes in Australia's climate. Pp. 35–51 *in* Greenhouse: planning for climate change ed by G. I. Pearman. CSIRO: Melbourne, and E. J. Brill: Leiden.

Raiffa, H., 1968. Decision analysis. Adddison-Wesley: Reading, Mass.

Spetzler, C. S. and Staël Von Holstein, C. S., 1975. Probability encoding in decision analysis. *Management Science* **22**: 340–57.

Valentine, H. T., Newton, C. M. and Talerico, R. L., 1976. Compatible systems and decision models for pest management. *Environ. Entomol.* **5**: 891–900.

Walters, C., 1986. Adaptive management of renewable resources. Macmillan: New York.

Whelan, R. and Main, A. R., 1979. Insect grazing and post-fire plant succession in south western Australian woodland. *Aust. J. Ecol.* **4**: 387–98.

PLATE 7

ABOVE: Revegetated minesite at Eneabba, Western Australia. A diverse array of shrub species has been re-established following sand mining, but does this represent the re-establishment of a functioning ecosystem? (Photo: R. J. Hobbs).

BELOW: A revegetated area in Tammin Shire, Western Australia. To what extent can revegetation aid in reconstructing functional ecosystems? What other components need to be reintroduced and how many species or functional groups are needed? (Photo: R. J. Hobbs).

CHAPTER 11

Is biodiversity important for ecosystem functioning? Implications for research and management

R. J. Hobbs[1]

Introduction

The chapters in this book examine biodiversity in the mediterranean ecosytems of southern Australia, particularly southwestern Australia, from a number of viewpoints. The aim of this chapter is to draw out the major conclusions and determine the implications of these for future research and management. As an aid to doing this, however, it became necessary to clarify the concepts discussed in the book, and for this purpose a simple conceptual model is developed which aims to provide a context for discussions of the role of biodiversity in ecosystem functioning in natural and human-modified systems.

A conceptual model

Primary ecosystem functions are the transfers of energy, nutrients and water (Hobbs, Chapter 1). It can be postulated that these functions can be carried out by a relatively small number of species. For instance a monoculture pine plantation performs these functions in a similar way to a diverse eucalypt forest (in that energy, nutrient and water transfers undoubtedly occur in each, although rates may differ). In other words, one hypothesis is that biodiversity is not important in ecosystem functioning. In fact, it could be argued that no clear link between diversity and function has yet been demonstrated (e.g., Vitousek and Hooper 1993), despite the fact that most biologists would assume that the two are interlinked.

---

[1]CSIRO, Division of Wildlife and Ecology, LMB 4, PO Midland, Western Australia, Australia 6056.

Pages 211–29 *in* BIODIVERSITY OF MEDITERRANEAN ECOSYSTEMS IN AUSTRALIA ed by Richard J. Hobbs. Surrey Beatty & Sons, 1992.

This problem has been discussed by O'Neill *et al.* (1986), who suggested that there was a basic distinction between the "population-community" or "biotic" approach to ecosystems and the "process-functional" approach. The former considers an ecosystem in terms of its constituent species, or populations, organized into guilds and communities, whereas the latter considers it in terms of the functions discussed above. O'Neill *et al.* (1986) state that "The problem is that the relationship between organisms in a system and the functions of the system is not always clear." The higher levels of organization that can be detected in a functional observation set do not always correspond in any simple manner to the levels of organization in the population observation set (MacMahon *et al.* 1978, 1981). Functional components need not be a single species, or even an aggregate of species (Webster 1979). Similarly, different species can perform the same or similar functions, the same species can perform several different functions at the same time, or the same species can perform different functions at different times or places (O'Neill 1986). This apparent mismatch may account for some of the difficulty experienced by the authors in this book when trying to grapple with the concept of the role of biodiversity in ecosystem function. The problem then becomes one of fitting organism-oriented concepts such as biodiversity into the process-functional ecosystem framework.

The solution lies in the breakdown of the overall ecosystem functions of energy, nutrient and water transfers into their constituents. These broad functions require a larger number of functions or processes which usually operate at the sub-ecosystem level. These functions or processes are predominantly biotic, and operate at the population or community level. It is thus at this level that biodiversity must be considered. For each function at this sub-ecosystem level, there will be a number of functional groups, each comprising a number of species (Fig. 1). This approach is essentially an application of hierarchy theory, as developed by Allen and Starr (1982), Allen *et al.* (1984), Pickett *et al.* (1989) and Allen and Hoekstra (1991); see also Huston *et al.* (1988). From this approach it becomes clearer that, while biodiversity may appear relatively unimportant at the broad level of ecosystem function, it may have a role by contributing to the various subfunctions.

This approach ties in well with the hierarchical approach to biodiversity I discussed in Chapter 1. Genetic diversity can be seen to operate at a further level of organization below that of the population and community, and thus feeds into the subfunction

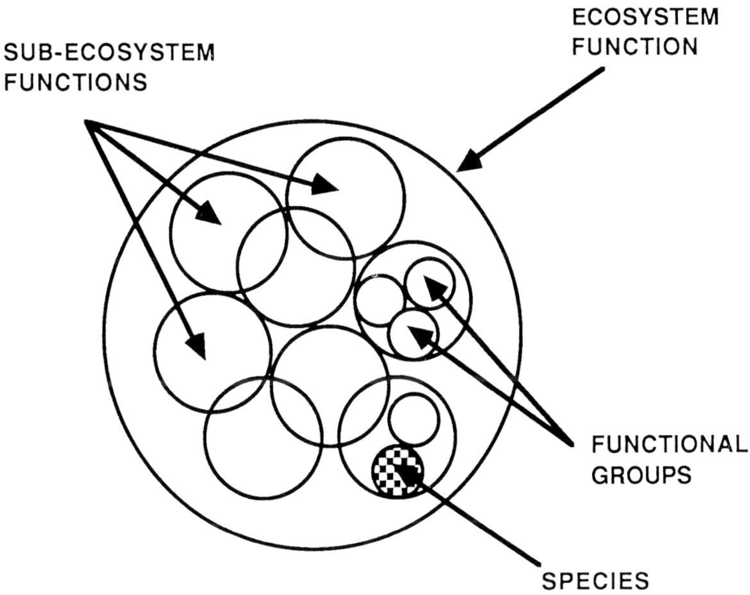

Fig. 1. Conceptual model illustrating a hierarchical approach to ecosystem functioning, illustrating how a primary function (nutrient, water or energy transfer) is made up of a number of different sub-ecosystem functions (e.g., decomposition, N fixation, species interactions, etc.), each of which is carried out by one or several functional groups consisting of groups of functionally equivalent species.

level up the hierarchy. Similarly, while we have not examined the importance of diversity at the landscape level in great detail in this book, the model in Figure 1 could be extended to include landscape processes as a higher level of complexity. Again, diversity becomes important at the subfunction level, but here additional important components of diversity may be ecosystem or landscape patch diversity (e.g., Urban *et al.* 1989; Kotliar and Wiens 1990).

Land use and modification of ecosystem function

Since mediterranean ecosystems in Australia are mostly heavily modified or fragmented, maintenance of biodiversity and ecosystem function in them will depend on effective management. In addition to past changes in land use, the current situation is very dynamic, with further likely changes in land use and possibly climate (Saunders and Hobbs, Chapter 4). An approach which encompasses this dynamism is thus required, although this was not enunciated in the preliminary treatment of the biodiversity issue (Solbrig 1991).

Using the above model, it becomes apparent that ecosystem management requires consideration of the various sub-ecosystem functions which together make up the major functions of water, energy and nutrient transfers. Main (Chapter 5) has argued that diversity is essential to allow continued functioning in the face of the large degree of variability found in mediterranean areas in terms of climate and disturbance. Important biotic functions in natural ecosystems in this environment include responses to low water and nutrient availability, high summer temperatures, fire and pathogens. However, since much of mediterranean Australia has been extensively modified by human activity, it is also useful to compare natural with human-modified systems. Here I contrast a natural ecosystem with an agricultural system. The major difference between the two is a dramatic simplification and reduction in the number of functional groups present in the agricultural system (see also Swift and Anderson 1993). In the natural system all the required subfunctions are carried out by the functional groups present, and thus there is considerable overlap between the major functions (Fig. 2a). Energy, water and nutrient transfers are thus closely interlinked.

In contrast, in the agricultural system prevalent in the mediterranean regions of Australia, the functional groups present carry out only a small proportion of the sub-functions involved in water, energy and nutrient transfers which were present in the natural system (Fig. 2b). Hence the three major functions are poorly performed and overlap less with each other. The reduced diversity of functional groups present means that there are fewer options or pathways for each function. The areas of non-overlap for each function thus represent pathways which are unavailable in the agricultural system because the required subfunctions are not being carried out. Thus, for instance, the predominance of annual crops and pastures means that energy capture is very inefficient, since there is no plant cover for half the year. Water uptake and evapotranspiration are reduced compared to perennial-dominated communities, since rooting patterns and growth periodicities are altered. There is no longer a diversity of rooting depths and modes which take up water from different soil layers and at different times of year. Nutrient transfers are also significantly different since the plants mostly lack specialized roots or symbionts, decomposer communities are greatly simplified and

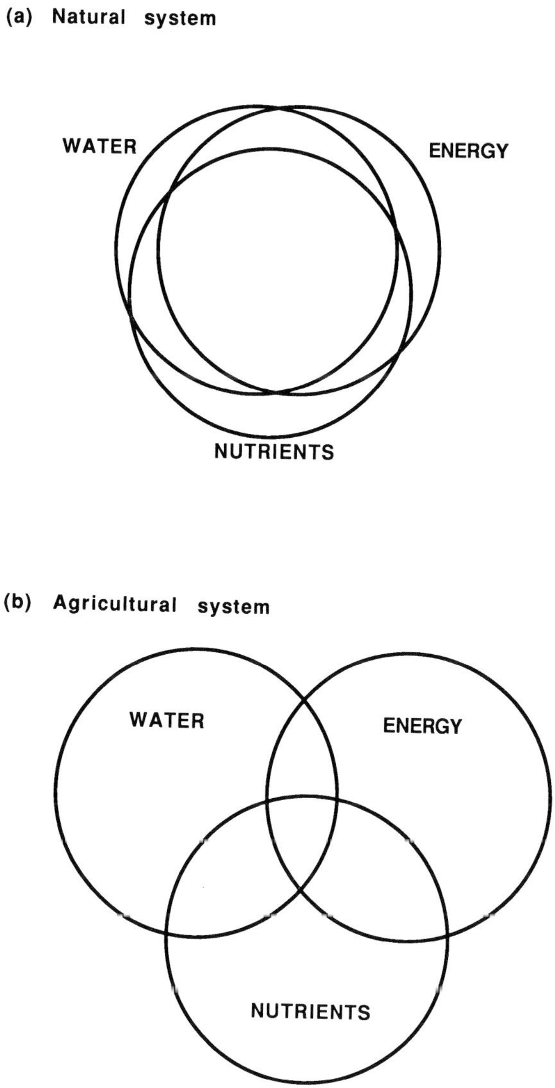

Fig. 2. Comparison of the major ecosystem functions in a natural system (a) *versus* a human-modified agricultural system (b). In the natural system, the functional groups and/or species present carry out sub-ecosystem functions such that the primary functions are maintained in relative balance. In the agricultural system, the transfers of energy, water and nutrients are not in balance since important functional groups are missing and hence sub-ecosystem functions are not carried out.

increased leaching and soil erosion lead to greater exports from the system (Lefroy *et al.* 1992; Hobbs *et al.* 1992). The agricultural system is thus very "leaky" compared with the natural system, and net flows of energy, water and nutrients in and out of the system are considerably greater (Hobbs 1993; Swift and Anderson 1993).

While the above analysis has considered the agricultural system, the same approach is valid for other human-modified systems. For instance, it is well established that timber extraction, especially in clear felling, leads to greater nutrient and water export from forest systems (Bormann and Likens 1981; Schofield *et al.* 1989). Mining activities such as bauxite removal in the jarrah *(Eucalyptus marginata)* forest and mineral sands extraction from the northern sandplains of southwestern Australia also completely disrupt ecosystem functions by removing all biota from the extraction site. However, in all these cases, these effects may result from the simple removal of structure and/or biomass rather than the reduction in diversity *per se.* Certainly, agriculture, logging and mining all result in a simplification of structure and composition, which results in reduced biodiversity (following the discussion of the definition of biodiversity in Hobbs, Chapter 1). The question remains whether the removal of individual species is important.

Little information is available on this question. The only example cited in this book which provides actual data on diversity versus function is that of Springett (1976), quoted by Lambeck (Chapter 7). In this study, soil microarthropod diversity and abundance and litter decomposition were compared in natural woodland and plantations of *Pinus pinaster* of three ages. Springett (1976) found no clear relationship between arthropod abundance and decomposition rates, but a significant correlation between species diversity and decomposition (Fig. 3). This figure indicates a large effect of increasing diversity at low levels of diversity, but this tails off at higher levels. This isolated example suggests an asymptotic relationship between diversity and function, corresponding to one of the models suggested by Vitousek and Hooper (1993). Thus a certain minimum number of species may be required for full ecosystem function, but the addition of more species may have little further effect on function (or rates thereof), except as insurance against disturbance and change, as suggested by Main (Chapter 5) and Lambeck (Chapter 7).

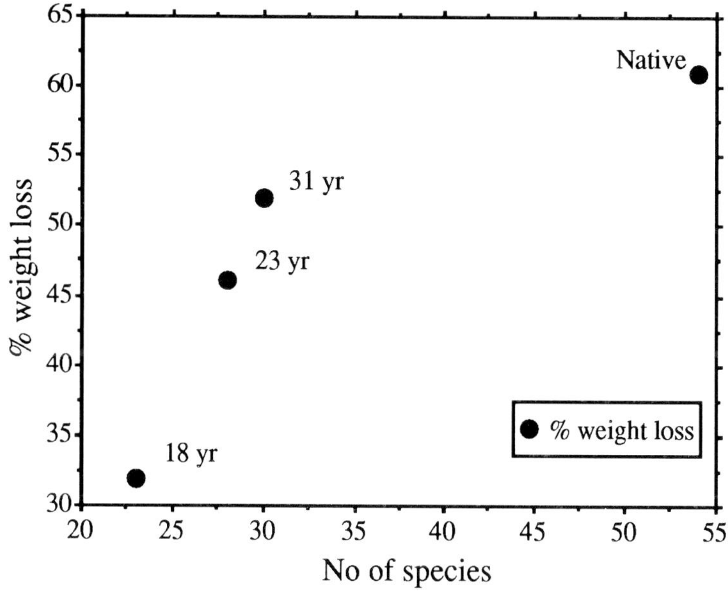

Fig. 3. Percentage weight loss (i.e., decomposition rate) from litter bags placed in plantations of *Pinus pinaster* of three ages (18, 23 and 31 years) and native woodland vegetation (predominantly *Banksia menziesii, B. attenuata* and *Eucalyptus todtiana* at Gnangara, Western Australia, *versus* number of species of microarthropods recovered from soil cores. Data from Springett (1976).

Management of ecosystem function

How does management fit into the scheme presented in Figure 2? In this context, the management aim becomes one of maximizing (or optimizing) the number of options available for the conduct of the three main functions. This should lead to a situation more akin to Figure 2a, in which each function has a full range of sub-functions and there is overlap between the functions; i.e., the functional groups present are capable of carrying out all the sub-functions necessary for all three functions. A system in which complete overlap occurred would have no net loss of water, nutrients or energy, and this is clearly impossible since no eco system is entirely self-contained. However, current management problems can be viewed as arising from a decreasing degree of overlap of the three functions as sub-ecosystem functions are lost or their rates modified.

The relative importance of different functions and sub-functions will depend on the land use under consideration: those sub-functions which are essential from a nature conservation perspective may be irrelevant from an agricultural perspective. In agricultural landscapes, the restoration of function (i.e., restoring hydrological balance or closing nutrient cycles) may be of prime importance, and this will entail re-introducing functional groups which allow these functions to be restored. For nature conservation, it is not the primary functions which are of primary interest. Rather, the sub-functions and the groups and/or species required to carry these out are important (see examples in Lamont, Chapter 6). Indeed, the retention of the species themselves is most frequently the objective of conservation management. In this case, therefore, the question becomes reversed. We are not interested in how species affect ecosystem function. Instead, we need to ask how functions have been affected by management or landscape modification. If it is an accepted goal that natural species and assemblages should be conserved in their own right, it does not matter whether we can prove that species diversity is important for ecosystem function. We need to emphasize instead the importance of the maintenance of ecosystem functioning for the conservation of species and ecosystems.

Modification to functions and sub-ecosystem functions in natural systems arises from losses of species and/or functional groups and additions of other species (invaders, pathogens, feral animals, etc.) on one hand, and physical and chemical changes brought about by ecosystem fragmentation on the other (Saunders *et al.* 1991; Hobbs 1993). Alteration of functions in remnants of native vegetation usually result from the changes occurring in the agricultural matrix, as discussed above. A characteristic of natural ecosystems is that they are self-replacing: modification of sub-functions may mean that self-replacement is no longer possible. For instance, regeneration of dominant tree species may be prevented by altered grazing pressures or fire regimes, changes in microclimate and soil characteristics, invasion by non-native plant species, or loss of pollinators or seed dispersers. Management for conservation thus requires that the key sub-ecosystem functions be recognized and manipulated as necessary.

Because of the degree of fragmentation and modification of natural ecosystems experienced in mediterranean Australia, there are now few, if any, areas which are capable of maintaining the type of functional integrity illustrated in Figure 2a. Individual reserves or remnants will not, in themselves, maintain ecosystem functions at "normal" rates because of the increasing influence of external factors arising in the surrounding heavily modified landscape. Because the situation in Figure 2b prevails in the surrounding production-oriented system, this will continue to force the natural system away from the state depicted in Figure 2a. This is due to the alteration in hydrologic regime over the whole landscape which resulted from the replacement of perennial vegetation with annual crops and pastures (McFarlane *et al.* 1992; Groves and Hobbs, Chapter 3), increased nutrient transfers in water and by wind erosion, and continual influx of non-native biota. Clearly, therefore, for the conservation of biodiversity within fragmented natural ecosystems to succeed, the management of surrounding lands must also be addressed (Hobbs and Saunders 1991; Saunders *et al.* 1991; Hobbs *et al.* 1992).

In the agricultural system, the major management problems arise because of poor performance of ecosystem functions, due to the absence of functional groups and hence important sub-ecosystem functions, as discussed above. Land degradation problems of salinization and erosion (see Saunders and Hobbs, Chapter 4) are directly related to the poor ability of the agricultural system to capture energy, use water and retain nutrients. Management of the agricultural system thus requires an increase in the numbers of functional groups which carry out the major functions. Optimally, the groups added should be of benefit to each function, hence leading to greater overlap, as in Figure 2a. An approach to this has been developed by Lefroy *et al.* (1992, 1993) which is based on increasing the amount of perennial vegetation in the agricultural landscape. The effect of this is to push the agricultural system back in the direction of the natural system and to tackle imbalances in energy, nutrient and water transfers simultaneously. The approach is thus one of increasing the complexity of the landscape (which can be viewed as increasing biodiversity at this scale) by reintroducing functional groups which had been removed during agricultural development. It remains to be seen whether the reintroduction of one set of functional groups (i.e., deep rooted perennials of varying types) will allow the establishment of other groups (e.g., decomposers) and the overall restoration of ecosystem function (Saunders *et al.* 1993). There are also numerous options available, ranging from the introduction of one perennial species, which could be native or non-native to the area, to an attempt to reconstruct a multi-species natural community with all structural components and functional groups present. The relative merits of each of these for the agricultural system will depend on how they affect the major ecosystem processes. However, if we are interested in enhancing the conservation value of the landscape too (i.e., through increasing habitat availability or providing corridors for movement), the choice of methods may have a significant impact (Lefroy *et al.* 1993).

From the above, it is clear that there is a need to integrate management for conservation and that for agriculture (or other production), and that management of individual landscape segments for one purpose without sympathetic management of surrounding segments will not work. There effectively needs to be an integrated approach to the management of natural and human-modified ecosystems such that the values of each are maintained or enhanced (Hobbs and Saunders 1991; Hobbs *et al.* 1993; Lefroy *et al.* 1993).

Research opportunities

Solbrig (1991) suggested that the key questions to be addressed in the study of the ecosystem function of biodiversity centred around our ability to predict what would happen following the removal or addition of individual species, or entire functional groups, structural groups or other ecosystem components. These questions could be addressed using three approaches. Firstly, controlled experiments involving the deliberate removal or addition of species or groups could be undertaken. Secondly, use could be made of inadvertent experiments involving loss or addition of species through human activities (such as fragmentation, exploitation). Thirdly, opportunistic use could be made of natural experiments where episodic events such as storms or droughts have altered species or functional group compositions.

It appears likely that the first option will be available only at very small scales, although these are still very important since adequate controls can be included. While the use of natural experiments can prove fruitful (e.g., Hobbs and Mooney 1992), the main source of information would seem to be from inadvertent experiments. In addition, a recurrent theme in this book has been the importance of recognizing the rapidity and extent of change which has occurred and is still occurring in Australian mediterranean ecosystems. Research cannot, therefore, be restricted to the few remaining "pristine" areas (if, indeed, any still exist), but must focus on the altered systems prevalent today. To be useful, research must examine questions relevant to the management and restoration of these altered systems in a dynamic and uncertain world.

Lavorel and Noble (Chapter 8) have illustrated how an integrated research approach is required which couples experiments, long-term observations and modelling. Other chapters in this book have indicated several possible areas for further study. Here I suggest some potential case studies which can yield useful information relevant to the question of the role of biodiversity in ecosystem function.

1. Removal of a functional group: Jarrah forest fire regimes

Nutrient cycling in the jarrah forest has been relatively well studied (Hingston *et al.* 1989). It is known that understorey leguminous species such as *Acacia pulchella* contribute significant amounts of nitrogen to the system post-fire through symbiotic fixation (Hansen *et al.* 1987, 1988). It is also known that these species establish from seed following fire due to heat-stimulation of germination, and that post-fire establishment could be manipulated by altering fire intensities, with low intensity fires and moist soil conditions reducing germination and establishment (Bell and Koch 1980; Shea *et al.* 1979; Bell *et al.* 1989; Portlock *et al.* 1990). In addition, it appears that production of the zoosporangial stage of the pathogen *Phytophthora cinnamomi* is significantly lower beneath leguminous stands than under other forest species (Shea *et al.* 1978; Murray *et al.* 1985). Leguminous species may either inhibit the pathogen directly, or via associated rhizosphere fungi (Whitfield *et al.* 1981; Malajczuk *et al.* 1984; Murray 1987; Dell and Malajczuk 1989)

Examination of nutrient cycling processes in plots where leguminous species abundance was modified experimentally could therefore yield useful information on the importance of this functional group to the functioning of the jarrah forest ecosystem. This can be supplemented with similar studies on minesite rehabilitation areas, where the role of understorey legumes has already been investigated in some detail (Ward *et al.* 1990, and see (3) below). In addition, identification of the causal mechanisms behind the reduction in pathogen abundance could identify whether the leguminous functional group also has a further role in pathogen suppression, either directly or through its fungal associates. Clearly, however, findings from such investigations should be placed in the appropriate context, given that today's jarrah forest is quite different to that present before extensive human modification (Dell and Havel 1989), and understorey legumes presumably may have played a different role in the past, perhaps as gap species.

2. *Removal of a functional group: effects of* Phytophthora cinnamomi

The invasive fungal pathogen is currently producing considerable floristic changes in some of the most important biological reserves in southwestern Australia, particularly the Stirling Range National Park (Wills 1992; Fig. 4). It also threatens to invade other important areas such as the Fitzgerald River National Park. It does not affect all elements of the flora to a similar extent, but has a greater impact on woody species, particularly members of the Proteaceae (Wills 1992). This results in a reduction in proteaceous cover and a concomitant increase in cover of herbaceous species, particularly Cyperaceae. Similar patterns of change have been found elsewhere (Weste 1986). There are two important aspects of this change which may affect ecosystem functioning. One is the effect of removal of Proteacea on nutrient uptake, given that this group has specialized root structures to enhance uptake (Lamont 1982, Chapter 6). The second is the removal of structural components of the ecosystem and of the nectar resources and seeds present in proteaceous flowers and fruits, which are important food supplies for a variety of bird, mammal and insect species (Wills 1992; see also Lambeck, Chapter 7, and Lamont, Chapter 6). Some of these species, in turn, are likely to be important pollinators. There is therefore the potential for studies which examine dieback affected areas in terms of nutrient regimes and utilization by fauna.

3. *Additions of species/functional groups in minesite rehabilitation*

Restoration ecology offers great opportunities for experimental ecology and testing ecological theories (Jordan *et al.* 1987). Rehabilitation of minesites following extraction of bauxite in the jarrah forest and mineral sands in the northern sandplain have received considerable amounts of ecological study (e.g., Bell *et al.* 1990; Ward *et al.* 1990). Ward *et al.* (1990) have illustrated the effects of the presence or absence of a leguminous understorey on the nitrogen economy of rehabilitation sites in the jarrah forest (cf (1) above). Majer (Chapter 9) has presented datasets on decomposition and on recolonization by Collembolan species in rehabilitation sites. While these datasets are not readily amenable to cross-comparison, it should be possible with further analysis to assess the effects of diversity of decomposer organisms with decompostion rates.

Fig. 4. (a) The Stirling Range National Park, a centre of floristic diversity in southwestern Australia, now under threat from the fungal pathogen *Phytophthora cinnamomi;* (b) and (c) *Banksia coccinea,* one of many species of Proteacea occurring in the Stirlings which is now threatened with extinction due to *Phytophthora* and introduced cankers. Photographs by R. J. Hobbs.

Bell *et al.* (1990) indicate the importance of ensuring that seeds of species with different seed storage strategies are returned to rehabilitation sites in the northern sandplain heaths. Reliance on seed stored in the soil ignores the importance of a large component of the flora which relies mostly on canopy storage. Even when these species are reintroduced by way of a mulching treatment, regeneration consists mainly of reseeding species. Few resprouters re-establish (S. Bellairs, pers. comm.). Thus there is a situation where a highly diverse community is re-established, but a major functional group (i.e., resprouters) is absent. This has important implications for the way the restored ecosystem responds to future disturbances. Experimental manipulation of rehabilitated minesites with different species/functional group complements could yield significant insights into the functional importance of various aspects of biodiversity.

4. Addition and removal of species/functional groups: fragmentation and revegetation in agricultural areas

Fragmentation of the Western Australian wheatbelt has resulted in a mosaic of patches of vegetation in varying conditions, from near "pristine" to severly degraded. Grazing, selective clearing, short-term cropping and other disturbances have left areas in which species or functional groups (of plants and/or animals) have been removed. On the other hand, invasions by non-native plants and animals have added species. There is considerable scope for investigating the functional impact of these changes.

In addition, as indicated above, problems of land degradation and loss of conservation value in agricultural areas need to be tackled in tandem. The solution currently being pursued is one of re-establishing perennial vegetation strategically across the landscape. It is relatively well established that doing this will alleviate many of the problems associated with rising water tables (i.e., salinization, erosion, waterlogging; McFarlane *et al.* 1992). Much of the revegetation work has been carried out using species not native to the region (for instance eucalypts from eastern Australia) but more recently there has been increasing impetus to use local species (e.g., Lefroy *et al.* 1991). While it seems likely that the species used will have little effect on the hydrological processes (except for relatively minor differences in water uptake rates), other processes and nature conservation value may be more closely linked to species composition. Studies such as those carried out by Springett (1976), comparing species diversities and ecosystem processes in areas dominated by native and non-native species in the forest region, need to be conducted in revegetated areas of the wheatbelt. If the revegetation is designed to act as habitat, food supply or movement corridors for fauna, then the types of species and functional groups used could determine the extent to which it is successful. To date, no studies have been carried out on the faunal use of revegetated areas, but this is an area which could yield useful information on the importance of diversity at a number of levels.

Conclusion

This book has presented a number of viewpoints on the topic of biodiversity and its role in the functioning of ecosystems in the mediterranean parts of Australia, particularly southwestern Australia. More questions have been raised than answers given. The variety of material presented indicates the breadth of the topic, and the different ways that it can be approached. The authors in general did not set out to provide a source book of data on the topic, but rather to explore the issues involved and produce a preliminary document to stimulate further discussion and research. While editing the book, I had to deal with conflicting referees reports on several occasions, with mixtures of praise and damnation for particular chapters. The topic lends itself to a considerable diversification of opinion, and we have to start somewhere with the process of refining our ideas. The book is, in effect, a "straw man" to be poked and prodded by others so that the ideas can be developed, or possibly rejected, as more information becomes available.

The functional significance of biodiversity is an important research topic but may not, in the end, prove to be a good reason for conserving biodiversity. We may find, for instance, that we cannot state categorically that all species present in an ecosystem are essential for its continued functioning. We should, perhaps, emphasize the importance of biotic diversity *per se* (i.e., its interest, beauty and intrinsic worth) rather than search for functional significance. As Lambeck (Chapter 7) points out, the ecological importance of biodiversity cannot be readily separated from these other values.

While it is still unclear whether the maintenance of biodiversity is essential for ecosystem functioning, there is no doubt that the maintenance of functioning ecosystems is essential if we want to retain biodiversity. Threats to biodiversity arise mostly from fragmentation and exploitation of natural areas, both of which disrupt or alter ecosystem functioning. The challenge for conservation is thus to find ways to maintian ecosystem processes in fragmented and extensively modified systems. We should not become fixated on searching for the functional significance of biodiversity and, in so doing, ignore the important and pressing problems relating to the maintenance of biodiversity in the face of massive changes in ecosystem processes.

Thus, although this is the end of the book, it is only the beginning of the process. However, time is short and we must ensure that we tackle the problems raised here in the most efficient way possible.

Acknowledgements

The ideas presented in this chapter arose from the discussion periods of the workshop, and I thank all the workshop participants for their contributions. I also thank Sandra Lavorel, Byron Lamont, Bert Main, Denis Saunders and Brian Walker for constructive and challenging criticism of the draft chapter.

References

Allen, T. F. H. and Hoekstra, T. W., 1991. Role of heterogeneity in scaling of ecological systems under analysis. Pp. 47–68 *in* Ecological heterogeneity ed by J. Kolasa and S. T. A. Pickett. Springer: New York.

Allen, T. F. H., O'Neill, R. V. and Hoekstra, T. W., 1984. Interlevel relations in ecological research and management: some working principles from hierarchy theory. USDA Forest Service General Technical Report RM-10.

Allen, T. F. H. and Starr, T. B., 1982. Hierarchy: perspectives for ecological complexity. University of Chicago Press: Chicago.

Bell, D. T. and Koch, J. M., 1980. Post-fire succession in the northern jarrah forest of Western Australia. *Aust. J. Ecol.* **5**: 9–14.

Bell, D. T., McCaw, W. L. and Burrows, N. D., 1989. Influence of fire on jarrah forest vegetation. Pp. 203–15 *in* The jarrah forest. A complex mediterranean ecosystem ed by B. Dell, J. J. Havel and N. Malajczuk. Kluwer: Dordrecht.

Bell, D. T., Vlahos, S. and Bellairs, S. M., 1990. Seed ecology in relation to reclamation: lessons from mined lands in Western Australia. *Proc. Ecol. Soc. Aust.* **16**: 531–35.

Bormann, F. H. and Likens, G. E., 1981. Pattern and process in a forested ecosystem. Springer: New York.

Dell, B. and Havel, J. J., 1989. The jarrah forest, an introduction. Pp. 1–10 *in* The jarrah forest. A complex mediterranean ecosystem ed by B. Dell, J. J. Havel and N. Malajczuk. Kluwer: Dordrecht.

Dell, B. and Malajczuk, N., 1989. Jarrah dieback — a disease caused by *Phytophthora cinnamomi*. Pp. 67–87 in The jarrah forest. A complex mediterranean ecosystem ed by B. Dell, J. J. Havel and N. Malajczuk. Kluwer: Dordrecht.

Hansen, A. P., Pate, J. S., Hansen, A. and Bell, D. T., 1987. Nitrogen economy in jarrah (*Eucalyptus marginata* Donn ex Sm.) forest of Western Australia. *J. Exp. Bot.* **38:** 26–41.

Hansen, A. P., Stoneman, G. L. and Bell, D. T., 1988. The potential of seeder legumes as a means to increase nitrogen availability in the jarrah forest. *Aust. For.* **51:** 226–31.

Hingston, F. J., O'Connell, A. M. and Grove, T. S., 1989. Nutrient cycling in jarrah forest. Pp. 155–77 *in* The jarrah forest. A complex mediterranean ecosystem ed by B. Dell, J. J. Havel and N. Malajczuk. Kluwer: Dordrecht.

Hobbs, R. J., 1993. Effects of landscape fragmentation on ecosystem processes in the Western Australian wheatbelt. *Biol. Conserv.* In press.

Hobbs, R. J. and Mooney, H. A., 1992. Effects of episodic rainfall events on Mediterranean-climate ecosystems. *In* Timescales of biological responses to water constraints ed by J. Roy, J. Aronson and F. di Castri. Springer: New York, in press.

Hobbs, R. J. and Saunders, D. A., 1991. Re-integrating fragmented landscapes — a preliminary framework for the Western Australian wheatbelt. *J. Env. Manag.* **33:** 161–67.

Hobbs, R. J., Saunders, D. A. and Arnold, G. W., 1993. Integrated landscape ecology — a Western Australian perspective. *Biol. Conserv.* In press.

Hobbs, R. J., Saunders, D. A., Lobry de Bruyn, L. A. and Main, A. R., 1992. Changes in biota. Pp. 65–106 *in* Reintegrating fragmented landscapes: towards sustainable production and nature conservation ed by R. J. Hobbs and D. A. Saunders. Springer: New York.

Huston, M., De Angelis, D. and Post, W., 1988. New computer models unify ecological theory. *BioScience* **38:** 682–91.

Jordan, W. R. III, Gilpin, M. E. and Aber, J. D., 1987. Restoration ecology: a synthetic approach to ecological research. Cambridge University Press: Cambridge.

Kotliar, N. B. and Wiens, J. A., 1990. Multiple scales of patchiness and patch structure: a hierarchical framework for the study of heterogeneity. *Oikos* **59:** 253–60.

Lamont, B., 1982. Mechanisms for enhancing nutrient uptake in plants, with particular reference to mediterranean South Africa and Western Australia. *Bot. Rev.* **48:** 597–689.

Lefroy, E. C., Hobbs, R. J. and Atkins, L. A., 1991. A revegetation guide to the central wheatbelt. Western Australian Department of Agriculture, CSIRO and Greening Australia: Perth.

Lefroy, E. C., Hobbs, R. J. and Scheltema, M., 1993. Reconciling agriculture and nature conservation: towards a restoration strategy for the Western Australian wheatbelt. *In* Nature conservation 3: reconstruction of fragmented ecosystems, global and regional perspectives ed by D. A. Saunders, R. J. Hobbs and P. R. Ehrlich. Surrey Beatty & Sons: Chipping Norton, NSW.

Lefroy, E. C., Salerian, J. and Hobbs, R. J., 1992. Integrating ecological and economic considerations: a theoretical framework. Pp. 209–44 *in* Reintegrating fragmented landscapes: towards sustainable production and nature conservation ed by R. J. Hobbs and D. A. Saunders. Springer: New York.

MacMahon, J. A., Phillips, D. L., Robinson, J. V. and Schimpf, D. J., 1978. Levels of biological organization: an organism-centered approach. *BioScience* **28:** 700–04.

MacMahon, J. A., Schimpf, D. J., Andessen, D. C., Smith, K. G. and Bayn, R. L., 1981. An organism-centered approach to some community and ecosystem concepts. *J. Theor. Biol.* **88:** 287–307.

McFarlane, D. J., George, R. J. and Farrington, P., 1992. Changes in the hydrologic cycle. Pp. 146–86 *in* Reintegrating fragmented landscapes: towards sustainable production and nature conservation ed by R. J. Hobbs and D. A. Saunders. Springer: New York, in press.

Malajczuk, N., Pearce, M. and Litchfield, R. T., 1984. Interactions between *Phytophthora cinnamomi* and *Rhizobium* isolates. *Trans. Br. Mycol. Soc.* **80:** 305–12.

Murray, D. I. L., 1987. Rhizosphere microorganisms from the jarrah forest of Western Australia and their effects on vegetative growth and sporulation in *Phytophthora cinnamomi. Aust. J. Bot.* **35:** 567–80.

Murray, D. I. L., Darling, D. D. and McGann, L. R., 1985. Indirect effect of florisitic composition on production of sporangia by *Phytophthora cinnamomi* in jarrah forest soils. *Aust. J. Bot.* **33:** 109–13.

O'Neill, R. V., 1976. Ecosystem persistence and heterotrophic regulation. *Ecology* **57:** 1244–53.

O'Neill, R. V., DeAngelis, D. L., Waide, J. B. and Allen, T. F. H., 1986. A hierarchical concept of ecosystems. Princeton University Press: Princeton, New Jersey.

Pickett, S. T. A., Kolasa, J., Armesto, J. J. and Collins, S. L., 1989. The ecological concept of disturbance and its expression at various hierarchical levels. *Oikos* **54:** 129–36.

Portlock, C. C., Shea, S. R., Majer, J. D. and Bell, D. T., 1990. Stimulation of germination of *Acacia pulchella*: laboratory basis for forest management options. *J. Appl. Ecol.* **27:** 319–24.

Saunders, D. A., Hobbs, R. J. and Ehrlich, P. R. (eds), 1993. Nature conservation 3: reconstruction of fragmented ecosystems, global and regional perspectives. Surrey Beatty & Sons: Chipping Norton, NSW.

Saunders, D. A., Hobbs, R. J. and Margules, C. R., 1991. Biological consequences of ecosystem fragmentation: a review. *Conserv. Biol.* **5:** 18–32.

Schofield, N. J., Stoneman, G. L. and Loh, I. C., 1989. Hydrology of the jarrah forest. Pp. 179–201 *in* The jarrah forest. A complex mediterranean ecosystem ed by B. Dell, J. J. Havel and N. Malajczuk. Kluwer: Dordrecht.

Shea, S. R., Gillen, K. J. and Kitt, R. J., 1978. Variation in sporangial production of *Phytophthora cinnamomi* Rands on jarrah (*E. marginata* Sm.) forest sites with different understorey compositions. *Aust. For. Res.* **8:** 219–26.

Shea, S. R., McCormick, J. and Portlock, C. C., 1979. The effects of fires in regeneration of leguminous species in the northern jarrah (*Eucalyptus marginata* Sm.) forest of Western Australia. *Aust. J. Ecol.* **4:** 195–205.

Solbrig, O. T., 1991. From genes to ecosystems: a research agenda for biodiversity. IUBS: Paris.

Springett, J. A., 1976. The effect of planting *Pinus pinaster* Ait. on populations of soil microarthropods and on litter decomposition at Gnangara, Western Australia. *Aust. J. Ecol.* **1:** 83–87.

Swift, M. J. and Anderson, J. M., 1993. Biodiversity and ecosystem function in agricultural systems. In press *in* Ecosystem function of biodiversity ed by E.-D. Schultze and H. A. Mooney. Springer-Verlag: Heidelberg.

Urban, D. L., O'Neill, R. V. and Shugart, H. H., 1987. Landscape ecology. A hierarchical perspective can help scientists understand spatial patterns. *BioScience* **37**: 119–26.

Vitousek, P. M. and Hooper, D. U., 1993. Biological diversity and terrestrial ecosystem biogeochemistry. In press *in* Ecosystem function of biodiversity ed by E.-D. Schultze and H. A. Mooney. Springer-Verlag: Heidelberg.

Ward, S. C., Koch, J. M. and Nichols, O. G., 1990. Bauxite mine rehabilitation in the Darling Range, Western Australia. *Proc. Ecol. Soc. Aust.* **16**: 557–65.

Webster, J. R., 1979. Hierarchical organization of ecosystems. Pp. 119–31 *in* Theoretical systems ecology ed by E. Halfon. Academic Press: New York.

Weste, G., 1986. Vegetation changes associated with the invasion by *Phytophthora cinnamomi* of defined plots in the Brisbane Ranges, Victoria, 1975–1985. *Aust. J. Bot.* **34**: 633–48.

Whitfield, F. B., Shea, S. R., Gillen, K. J. and Shaw, K. J., 1981. Volatile compounds from the roots of *Acacia pulchella* R.Br. and their effect on *Phytophthora cinnamomi* Rands. *Aust. J. Bot.* **29**: 195–208.

Wills, R. T., 1992. The ecological impact of *Phytophthora cinnamomi* in the Stirling Range National Park, Western Australia. *Aust. J. Ecol.* in press.

PLATE 8

ABOVE: *Verticordia picta* and *Verticordia eriocephala* in kwongan vegetation in Durokoppin Nature Reserve in the Western Australian wheatbelt. (Photo: R. J. Hobbs).

BELOW: *Calytrix strigosa* in kwongan on Durokoppin Nature Reserve. (Photo: L. A. Atkins).

Index

F

Q

R

S